Teletechnologies, Place, and Community

Comedia

SERIES EDITOR: DAVID MORLEY

Teletechnologies, Place, and Community

Rowan Wilken

Routledge
Taylor & Francis Group

NEW YORK AND LONDON

First published 2011
by Routledge
605 Third Avenue, New York, NY 10017

Simultaneously published in the UK
by Routledge
4 Park Square, Milton Park, Abingdon, Oxon OX14 4RN

*Routledge is an imprint of the Taylor & Francis Group,
an informa business*

First issued in paperback 2014

The right of Rowan Wilken to be identified as author of this work has been asserted by him/her in accordance with sections 77 and 78 of the Copyright, Designs and Patents Act 1988.

Typeset in Sabon by IBT Global.

Library of Congress Cataloging-in-Publication Data
Wilken, Rowan.
   Teletechnologies, place, and community / by Rowan Wilken.
      p. cm. — (Comedia)
   Includes bibliographical references and index.
   1. Information technology—Social aspects.   2. Telematics—Social aspects.   3. Cyberspace.   4. Telecommunication—Social aspects.
   5. Electronic villages (Computer networks)   6. Communication and technology.   7. Architecture and technology.   I. Title.
   HM851.W55155 2011
   303.48'33—dc22
   2011004441

ISBN: 978-0-415-87595-0 (hbk)
ISBN: 978-1-138-77676-0 (pbk)

For Lazarus and Maxim and Sunday,
each of whom came at just the right time
and
For Dad, who went too soon but who always believed that he
was going home to a better place and perfect community

Contents

Figures

Acknowledgments

Many people contributed to the successful completion of this book. In particular, I would like to acknowledge the following: Bharat Dave, Isabelle d'Solier, Gerard Goggin, Lisa Gye, Sandra Kaji-O'Grady, Ramon Lobato, Christian McCrea, Scott McQuire, Esther Milne, Buck Clifford Rosenberg, John Sinclair, Darren Tofts, Andrew Vincent, Erica Wetter, and Geoff and Amanda Wild.

A special thank you goes to John Olsen for his meticulous reading of the manuscript (twice!), and to David Morley, who actively supported this project from the outset and who was instrumental in getting it to this point.

For their love, encouragement, and support in so many different ways, I thank my family: Shelagh Wilken, Lois Wilken, John and Shirley Olsen, and Geoff Olsen and Suzie Mitchell. Also and especially, I thank Karen Olsen for everything; you are the love of my life, and your assistance and encouragement have been invaluable.

These people, and many others, know the true significance of their support for my work. My sincere thanks go to them all.

I would also like to acknowledge the financial support provided by the Faculty of Life and Social Sciences, Swinburne University of Technology, in covering image reproduction costs, and the Faculty Teaching Support Fellowships Scheme at the University of Melbourne, which gave me some time away from teaching during my time there to work on the manuscript.

Parts of this book have been published elsewhere in the following journals, and these parts are reproduced here with permission. Portions of Chapter 3 appeared as 'The Haunting Affect of Place in the Discourse of the Virtual', *Ethics, Place and Environment: A Journal of Philosophy and Geography*, 10(1), March (2007): 49–63. One section of Chapter 4 appeared as 'Calculated Uncertainty: Computers, Chance Encounters, and "Community" in the Work of Cedric Price', *Transformations*, 14 (2007), <http://transformations.cqu.edu.au/journal/issue_14/article_04.shtml>. And, portions of Chapter 6 appeared as 'From *Stabilitas Loci* to *Mobilitas Loci*: Networked Mobility and the Transformation of Place', *Fibreculture Journal*, 6 (2005), <http://six.fibreculturejournal.org/>.

Introduction

Teletechnologies—or technologies of distance—cannot be ignored. The electronic age is said to have wrought profound changes to how we think about and experience who we are, where we are, and how we relate with one another. A rapidly changing and multilayered techno-social milieu has emerged, one that is characterised by complex interactions and interconnections between information and telecommunications technologies, the places in which we live, and various forms of social engagement or community.

The concepts of place and community have traditionally formed key frameworks for thinking about these issues, yet both are considered to have been transformed by networked teletechnologies. For instance, it has been suggested that, 'The worldwide computer network—the electronic agora—subverts, displaces, and radically redefines our notions of gathering, place, community, and urban life' (Mitchell, 1995: 8). Similar arguments have been made elsewhere, where it is argued that electronic media 'lead to a nearly total dissociation of physical place and social "place"' so that 'where we are physically no longer determines where and who we are socially' (Meyrowitz, 1985: 115).

The above sentiments highlight the complex and at times contradictory ways that place and community appear to circulate in relation to information and communication technologies. This is hardly surprising in light of the contradictions and complexities that also attend notions of place and community beyond computer-mediated communications and networked computing—such as in architecture, urban design, and studies of the domestic home, to name a few. Given this context, when, where, and how do concepts of place and community fit into ideas about a world transformed by teletechnologies?

The purpose of this book is to answer this question by examining the relationships between teletechnologies, place, and community. Specifically, it traces how these notions of place and community are employed in the writing on teletechnologies—principally in a specific period from the late 1980s to the early 2000s—and how these notions might be influenced by these technologies. After all, one of the great ironies of the information age—the age that was said to spell the death of the book—is that it has produced a vast archive of printed material devoted to documenting and explaining the changes that this age has wrought.

In respect to place, for example, in the academic and popular literature of this period, this is a concept which holds an ambiguous, almost double, presence. One the one hand, it is argued that networked computing technologies facilitate forms of space and place that 'may be unique to new media' (Manovich, 2001)—a development that leads the Dutch architect Rem Koolhaas to remark that, 'the only space that matters in our time is the one we connect to on the computer' (quoted in Knack, 2000: 9). Characterising many of the earliest accounts of this period was a particularly strong techno-boosterist rhetoric that conceived of this 'new media space' as 'dematerialised' and free from the constraints of geography and physics. On the other hand, and at the same time, place-based metaphors emerge as critical during this time to the conceptualisation and representation of life on-line.

Community, too, is a notion that circulates in the wider literature of the same period in equally complex ways. At the same time as it is viewed by some social commentators as a type of social form in significant wider decline, and viewed by others as a concept in need of radical revision, community re-emerges as 'new' forms of 'computer-mediated' or 'virtual' community enabled and facilitated by networked communications technologies.

This is just one example or snapshot of the aforementioned contradictions and complexities, and the aim of the book is to examine in what ways the literature on these topics contributes to how teletechnology, place, and community, are constructed, conceived of, and debated—both in academic as well as wider social discourse.

From this examination, a fuller, critical understanding of the interactions and interconnections between teletechnologies and notions of place and community is developed, in order to determine what significance these three elements have for understanding and living in the present digital age. If place and community are of any relevance, these notions must be understood in conjunction with our uses of information and telecommunications technologies, as part of a wider techno-social milieu.

In pursuit of these aims, this project is historiographical in approach, with a particular concern for the workings of discourse (in the specific sense of the language used to describe and debate new media technologies). It is also concertedly interdisciplinary. Each of these methodological concerns, or orientations, is elaborated below.

HISTORIOGRAPHY

In the introduction to the collection *New Media, 1740–1915*, Geoffrey Pingree and Lisa Gitelman (2003: xi) note how, 'In the short space of a current college [or university] student's lifetime, the internet has gone from a specialized, futuristic system to the network that most significantly structures how we engage daily with the world at large'. *Teletechnologies, Place and*

Community addresses itself to both these historical moments in time. That is, it looks backwards, to examine in detail the emergence of a flowering of popular and critical interest in the social and spatial possibilities associated with the Internet, while at the same time looking forward to the present (and future) by examining the ongoing implications of this earlier phase of engagement with teletechnologies, and concepts of place and community.

In light of this, this book is not only a work of interdisciplinarity, it is also a work of media history—or, more precisely perhaps, given its focus on the *writing of the history of* teletechnologies, like the Internet and computer-mediated communication (CMC), it is a work of media historiography.

The particular approach to 'media history' that is pursued here is informed by the work of film scholar Rick Altman (2004) and his concept of 'crisis historiography'. This is the idea that, while 'all media were once new' (Gitelman, 2008: 1), the initial conditions of 'newness' are never (pre-) settled. Rather, they are always negotiated and renegotiated, and thus 'the definition of a representational technology is both historically and socially contingent' (Altman, 2004: 16). 'Crisis historiography', then, is viewed as an approach to media history scholarship that is informed by the understanding 'that new media, when they first emerge, pass through a phase of identity crisis, a crisis precipitated at least by the uncertain status of the given medium in relation to established, known media and their functions' (Pingree and Gitelman, 2003: xii). According to Altman, this 'crisis of identity' is 'reflected in every aspect of [a] new technology's socially defined existence' (2004: 19). More specifically, Altman argues, this involves three processes which he summarises as 'multiple identities', 'jusidictional conflicts', and 'overdetermined solutions'.

In respect to the first of these, Altman writes that 'new technologies are always born nameless' and thus 'begin life with multiple monikers rather than a single stable name' (19). We see this, for example (as will be explored in more detail in later chapters), in the many semantic arguments over terms such as 'cyberspace', 'virtual community', and 'computer-mediated communication'.

With these multiple identities and attendant uncertainties comes conflict over who will establish control of and over the new medium. As Altman (2004: 20) observes, 'attempts to establish jurisdiction over new technologies extend broadly'. They manifest themselves in considerations of political economy. They are also manifest in representational terms. For instance, using the example of cinema to illustrate this point, Altman notes how 'individual films reflect the influence of multiple different media, thus replicating in miniature the overall conflict' (20)—so, for example, a film might combine 'deep-focus realism and a painted theatrical backdrop, [. . .] or a strong realist narrative and an illusion-breaking look directly into the camera' (20). In addition, and of particular interest in the context of this book, is the way that these jurisdictional disputes also manifest themselves discursively, through the ways that new media are written about and debated.

Finally, by 'overdetermined solutions', Altman is referring to the process by which, eventually, 'new technologies adopt a familiar name, identity, and features' (21). The crucial point here is that, 'far from happening passively or automatically, these overdetermined solutions are an indirect result of jurisdictional conflicts waged by users' (21). One key implication of this is that new technologies must be understood as 'socially constructed' in ways that are 'both ongoing and multiple' (21).

This last point helps to explain, at least in part, why it is that many of the key terms or tropes associated with the Internet literature under examination here (especially 'cyberspace' and 'virtual community') had their own particular currency and richness of meaning in a particular period, constituting forms of 'unfamiliar' language both prior to and subsequent to this period. Thus, in more recent new media scholarship, the notion of 'virtual community' has given way to (related) analyses of 'social networking', and the once popular 'cyberspace' has been supplanted by more generic descriptive terms, such as the Internet or world wide web (or, more recently still, the somewhat sarcastic colloquial term, 'interwebs', which was first 'coined' by George W. Bush); the same can even be said of more specialised terms, such as 'computer-mediated communication'. Nevertheless, given the commitment here to considering these specific teletechnological (and associated social) forms 'within their own historical contexts' and in ways that 'seek out the past on its own passed terms' (Pingree and Gitelman, 2003: xi), this earlier language is consciously retained and employed here.

The value of the 'crisis historiographical' approach for the present project is thus usefully summarised by Gitelman as follows: 'Looking into the novelty years, transitional states, and identity crises of different media stands to tell us much, both about the course of media history and about the broad conditions by which media and communication are and have been shaped' (2008: 1). So, to take one example, while this book addresses 'virtual community', but not the shift that has taken place in the space of only a few years to 'social networking' (Facebook, Twitter, etc.), focusing on these earlier technological and social forms can reveal much about their more recent (re)incarnations.

DISCOURSE AND TEXTUAL CRITIQUE

The Australian expatriate media theorist McKenzie Wark has quipped that, 'The growth industry of the '90s [was] not multimedia, cyberspace or virtual reality. The growth industry of the '90s [was] *hype* about multimedia, cyberspace and virtual reality' (Wark, 1997: 155). In the present context, the above remark carries twofold significance: it reinforces the arguments set out above about the need to historicise new media; it also gestures towards the importance of paying due attention to the *language* that is used to describe and debate new media, which is one of the key aims of this book.

Teletechnologies, Place, and Community asks how notions of 'place' and 'community' circulate and are mobilised within the literature on teletechnologies. What discursive strategies are employed? How are 'place' and 'community' used metaphorically, and what drives the selection and use of these particular figures of speech? What is buried within them? What 'aporias' or 'critical blind-spots' (Lucy, 2004: 1) might be revealed through their usage and through the usage of other rhetorical and discursive strategies? Moreover, what are the implications of these choices and strategies? How do they shape the way that teletechnologies, place and community are imagined, conceived, and debated past, present, and in the future?

Such questioning, which involves an 'opening of media analysis to literary, hermeneutic, and rhetorical forms of study' (Real, 1996: 120), is productive insofar as 'texts employ specific forms and conventions to give shape and purpose to our media experience' (120), and close textual analysis of these texts can explore and reveal those devices. It is in this sense that the book develops a close *critique* of the available literature on teletechnologies, as this literature involves notions of place and community, with the term critique understood in a very precise sense:

> A critique of any theoretical system is not an examination of its flaws or imperfections. It is not a set of criticisms designed to make the system better. It is an analysis that focuses on the grounds of that system's possibility. The critique reads backwards from what seems natural, obvious, self-evident, or universal, in order to show that these things have their history, their reasons for being the way they are, their effects on what follows from them, and that the starting point is not a (natural) given but a (cultural) construct, usually blind to itself. (Johnson, 1981: xv)

As should be clear from the above passage (Johnson's quote is drawn from her introduction to Derrida's *Dissemination*), this particular approach to textual critique and how it is applied in this book is informed strongly by the work of Derrida. It also draws on Althusser and Balibar's (1970: 25–28) notion of 'the problematic'—which Storey (2001: 96) summarises as the critical role of attending to the 'assumptions, motivations, underlying ideas, etc.' from which any cultural text is made. More particularly, though, the specific understanding of textual critique developed and applied here also takes inspiration from the literary theory of Pierre Macherey.

Drawing from Freudian psychoanalysis, Macherey (2006) contends that 'the act of knowing' what goes on in a text involves the 'articulation of a silence' (6), or what he calls a text's 'unconscious'. Thus, he writes, 'to explain [a] work is to show that [. . .] it is not independent, but bears in its material substance the imprint of a determinate absence, which is also the principle of its identity' (89). Viewed in this way, Storey suggests that:

> The task of a fully competent critical practice is not to make a whisper audible, nor to complete what the text leaves unsaid, but to produce a new knowledge of the text: one that explains the ideological necessity of its silences, its absences, its structuring incompleteness—the *staging* of that which it cannot speak. (Storey, 2001: 100)

In short, Macherey's contention is that, whenever something is said, something else is left unsaid. 'What is important in the work,' he writes, 'is what it does not say' (2006: 97), and, therefore, 'we investigate the silence, for it is the silence that is doing the speaking' (96). In this way, I agree with Storey (2001) that, 'it is the reason(s) for these absences, these silences, within a text which must be interrogated' (101). These are crucial concerns in the examination of teletechnologies, place, and community that follows, where it is asked, to cite one example, why so many of the early accounts of 'cyberspace' and 'virtual community' employ place-based metaphors to provide a temporary sense of 'spatiality' to on-line social interaction, but where these metaphors are ultimately considered extrinsic or supplementary to this interaction, and in some cases actively resisted. One argument that is developed in this book is that 'place' is one of the 'silences' that 'speaks' in this literature.

INTERDISCIPLINARITY

In the context of the present endeavour, my concern for interdisciplinary research has been generated by two factors. First, in contextual terms, I have, over the course of this project, moved between the disciplinary setting of media, literature and film, to architecture, to Australian studies, to cultural studies, and back to media. This itinerant and varied research and employment trajectory has brought many fresh perspectives to my work, as well as an equal number of challenges in explaining the intricacies of this project, and has confirmed and reconfirmed for me the need to be able to 'listen' as well as 'speak' to (and across) these (and other) disciplinary contexts. Secondly, interdisciplinarity emerged very early in this project as a methodological necessity. As already stated, a primary task of this book is an examination of the interconnections between teletechnologies, place, and community. In order to consider these three things in combination, and to do full justice to an examination of the complexities around them, the project draws from the discourses of a number of different disciplines, including media and communications, philosophy, geography, sociology, architecture, urban planning, literature, and more. This is necessary insofar as any account of how notions of place and community are to be understood in relation to teletechnologies, has to also understand the wider contexts in which these terms circulate and are debated.

While currently quite a fashionable term, interdisciplinarity is nevertheless something of a contentious concept. For instance, Alan Liu has famously described interdisciplinary study as 'the most seriously underthought critical, pedagogical and institutional concept in the modern academy' (Liu, 1989: 743). Liu's point is that, while fashionable, interdisciplinary studies can often serve either as a 'title conferred upon [a] wonder cabinet of ill-sorted methods' (743), or a term that masks the fact that what is meant in practice by interdisciplinary study frequently remains unclear.

In addition to the above criticisms, there is also considerable variance in how the concept of interdisciplinary research is defined. For instance, in his book *Into the Light of Things: Art of the Commonplace from Wordsworth to John Cage* (1994), George J. Leonard, Professor of Interdisciplinary Humanities at San Francisco State University, defines interdisciplinary work as 'at heart the urge to break artificial bounds' (xiii). The interdisciplinary researcher, Leonard argues, 'consciously reconnects what has been artificially divided' (xiv).

While respecting Leonard's definition, this book acknowledges the 'artificiality' of conceptual and disciplinary boundaries, but nevertheless also acknowledges their ongoing existence and maintenance and the influence these boundaries bring to bear on the formation and direction of knowledge. For this reason, the conception of interdisciplinarity mobilised here has greater affinity with the position of the Australian media theorist Andrew Murphie (2004) who describes interdisciplinary research as very much 'a matter of meetings of differences'. According to this view, it is important *not* to see these differences as the result of entrenched lines of division between disciplines—of what is or is not a given discipline's purview. Rather, a more productive approach is to see such disciplinary difference, to use Doreen Massey's words, as the result of differing 'angles of approach' (1999: 6) to particular issues or key concepts (such as 'space', to use the example of architecture and media).[1]

In a book-length study of the concept of interdisciplinarity, Joe Moran (2002) also makes the broader point that, while one approach is to think of this concept as involving the quest for 'total knowledge' (15), the prevailing approach to interdisciplinarity 'tends to be centred around problems and issues that cannot be addressed or solved within the existing disciplines' (15) and which require an outward looking focus.

Moran lays out three further and interconnected points about interdisciplinarity which bear description here, as they are central to why interdisciplinary research is viewed as a particularly productive approach in the context of this project. Moran's first point is that we 'cannot understand interdisciplinarity without first examining the existing disciplines' (2).[2] Embedded in this first, baseline understanding of interdisciplinary research is an acknowledgment of the labour involved, insofar as such research is premised on, and can only proceed from, a thorough examination of existing disciplinary knowledge. Building on this, Moran's second point is that interdisciplinarity functions as a 'reflexive form of disciplinarity that recognises its own limits' (187). In other words, interdisciplinary study, for Moran, 'represents, above all, a denaturalization of knowledge' (187) which, in turn, promotes a more critical and self-conscious engagement with existing disciplines and disciplinary knowledge, including one's own. His third point is that, through interdisciplinary research, our 'most basic assumptions can always be challenged or reinvigorated by new ways of thinking from elsewhere' (187). The significance of this is that interdisciplinary research is never just a 'documentary' process of acknowledging or 'recording' the existence of alternative disciplinary forms of knowledge; rather, it is always a creative and productive process which leads to deepened knowledge and reconfigured ways of thinking.

The implications of the above for the present project are threefold. First, the need to be attentive to and respectful of differing angles of disciplinary approach is recognised. Secondly, and at the same time, there is also a need to develop, at least provisionally, a 'shared language' (Massey, 1999: 5) that enables conversation 'across the boundaries' (5) of these disciplines while being respectful of the traditions that contribute to conceptual differences. Nevertheless, thirdly, it is imperative that the assumptions, ideologies, and other forces which structure and motivate these differing angles of approach be interrogated.

Having framed the importance of interdisciplinarity to this project, it is perhaps prudent at this point, especially given the earlier arguments regarding discursive silences, to acknowledge at least one key 'silence' of the present text. This is important in light of Gitelman's (2008: 1) observation that media history is inevitably always partial in that it takes on particular foci depending on the methods and approaches employed.

The explicit focus of this book is on European and Anglo-American cultural contexts, and with teletechnologies as they have been discussed within English-language written texts—sources, it should be remembered, which are generated, in the main, by academics and commentators publishing within Australian, North American, and European publishing contexts and markets. While it has not been intended, this double focus imposes a number of restrictions on the present study. Admittedly, these restrictions are helpful insofar as they limit and focus the scope of the present study. But this same limit also means that what is developed here is one particular take on the issue of teletechnology, place, and community, and it has the potential to exclude or overlook other, alternative approaches to the topic, including calls to 'internationalize internet studies' and to move beyond Anglophone paradigms (Goggin and McLelland, 2009a). While it is beyond the scope of this book to engage with this issue in detail, it is one that will be revisited towards the end of the book.

Having provided this qualification, and prior to detailing the structure of the book, two brief remarks need to be made regarding use and meaning of each key term.

The first concerns the use of the term 'teletechnology'. Writing on the word 'television', Raymond Williams offers the following remarks:

> The central technical claim of television is that it can show us distant events. The hybrid name selects this quality, following *telescope, telegraph, telephone, telepathy*, with *tele* as the combining form, from the Greek for 'afar', related to *telos*, 'end'. (Williams, 1989: 13–14)

Prefiguring many of the debates about the Internet and distance (which are to be taken up in the next chapter), Williams views the experience of 'telepresence' (presence at a distance) in relation to television as self-evident, 'usually obvious':

> We are in one place, usually at home, watching something in another place: at variable distances, which however do not ordinarily matter, since the technology closes the gap to a familiar connection. (1989: 14)

However, what is sometimes perhaps less obvious, but no less important, is the fact that, although 'this kind of distance is the essence of tele-vision (seeing from afar) [. . .] this should not be understood as some kind of technological inevitability' (O'Connor, 1989: 92). This is to say that, while 'tele'-effects of all sorts may well create 'a familiar connection', 'an illusion' (Williams, 1989: 14), these effects are made (not natural); they are constructed. In this way, like the textual 'silences' noted earlier, remaining attentive to what the implications of these 'tele-effects' might be is a central (if at times implicit) preoccupation of this book.

For these reasons, the term 'teletechnologies' is preferred here over other terms, such as information and communications technologies (ICTs), computer-mediated communications (CMC), computer-supported social interactions (CSSI), and others. Teletechnology is more compact than these and other expressions, as well as being more readable than their acronyms. It is also more suggestive, insofar as it holds within the prefix 'tele' this very idea— this very important idea—of 'technologies of distance'. Distance is a central concern in much of the established writing on the development of new media (Mitchell, 1995; Meyrowitz, 1985; Martin, 1981), and the word teletechnology retains this sense of technologies of distance at the very same time as this idea is to be both tested and critiqued—especially around the notions of place and community. In other words, the word 'teletechnology' operates as a kind of shorthand, or metonym, for many of the concerns under investigation in this book. Having said this, other established terms within the literature, such as ICT and CMC, will still be referred to where pertinent.

The other two key terms—'community' and 'place'—are treated at length in Chapters 1, 2, and 3. However it is worth noting here the use of the singular throughout this book when referring to 'place' and 'community'. This is because the immediate concern is in tracing how the *idea* or *notion* of 'place' and of 'community' operates in the discourses and available documentary sources concerning teletechnologies. Even so, this usage should not be taken to represent an argument for the monolithic or the abstract over the multiple (place*s* and communit*ies*) or the particular. The latter understanding is both implicit and assumed throughout, and is more explicitly argued for in the final chapter.

And so, finally, to an explanation of the structure of this book which examines three core areas or themes: teletechnologies, place, and community, and then the points of intersection overlapping between them.

Chapter 1 sets up a context for this project by tracing the connections between ideas of community and teletechnology use in the social sciences and media and communications literature. Specifically, it examines the rise of the notion of 'virtual community'—a term understood to refer to particular forms of social interaction and aggregation made possible by computer-mediated communication technologies. Discussion traces the coining of the term virtual community, the rapid proliferation of its popular usage, and the various evolutions in its critical application. Detailed consideration is also given to what is at stake in, and the issues raised by, computer-mediated communication exchange and research.

From this analysis of virtual community, Chapter 2 necessitates a step back from teletechnology use, to concentrate more closely on the history of the concept of 'community', which has had such prevalent use as a metaphor in writing on teletechnologies. The aim of this chapter is to first grasp and then attempt to respond to some of the problems that are understood by many scholars to attend the notion of community. It does this by exploring three different sets of perspectives on the concept of community that have been developed over the course of the twentieth century.

In Chapter 3, emphasis shifts from considerations of community to that of computer-mediated communication and place. The chapter begins with a consideration of the concept of place, and proceeds to examine how the notion of place circulates in relation to teletechnology use and virtual community. Here the examination draws from a range of disciplines, including geography, media and communications, philosophy, and architectural theory to explore how the concept of place functions as a metaphor in writing on teletechnologies, and what the implications are of these metaphorical conceptions of place.

Chapters 4 and 5 both focus on the points of intersection between teletechnology use and place at different points in the history of architectural writing on computer-aided design.

Chapter 4 examines architecture's early engagement with computing technologies. The concern of this chapter is with tracing this interest in computing technologies, and with understanding how it is married to a concern for questions of place and community in the architectural literature. The primary aim of this chapter is to understand how an explicit engagement with computers might be informed by, and in turn shape, ideas of place and community.

These concerns are further developed in Chapter 5, which gives detailed consideration to mapping the sorts of ideas which motivate more recent architectural interest in computers and cyberspace in order to understand how this differs from early use of computing technologies in architecture and urban design. As with the previous chapter, the aim is to examine the extent to which such an explicit engagement with computing technologies both shapes and is shaped by ideas of place and community. This focus also provides a useful basis for comparison with the findings of Chapter 1 and 3 and their examinations of place and community in writing on computer-mediated communication and cyberspace.

Chapter 6 moves from the abstractions of cyberspace architecture to the more quotidian setting of the domestic house, a key site in the 'naturalisation' and consumption of computing and other communications technologies. Within the literature, the home is widely understood as a complex site which incorporates various teletechnologies and is shaped by the interplay and tensions of public and private, global and local. These claims are examined and tested against some historical understandings of home. In the second half of the chapter, these developments and understandings are then contrasted with the rise of networked (especially telephonic) mobility. A key overall aim of the chapter is to discover how we might begin to make sense of rapid developments in domestic and mobile technologies. A further aim is to understand how domestic and mobile technologies, and approaches to thinking about them, might impact on and shape understandings of place and community.

Finally, in light of the material of the preceding chapters, Chapter 7 returns to the three core concepts explored in this book to develop a three-part proposal that explores how we might best engage with teletechnologies in concert with suitable, reworked understandings of place and community.

Having provided this summary of the aims, methods, and overall terrain of the book, it is time to turn now to Chapter 1 and an examination of computer-mediated communication and 'virtual community'.

1 Techno-Sociality
Computer-Mediated Communication and Virtual Community

'The familiar will always remain the likely starting point for the rendering of the unfamiliar.'

—E. H. Gombrich (1972: 72)

Gombrich's aphorism is especially pertinent in the context of early Internet scholarship which sought to make sense of the social possibilities presented by the emergence of networked computing technologies. In the main, it did this by redeploying the familiar concept of 'community' as 'virtual community' in order to encapsulate the less familiar social formations that were emerging through computer-mediated social interaction. It is this concept of 'virtual community' that is the principal focus of this chapter. The aim of this chapter is to develop a deeper understanding of this notion and how it circulates in the academic and popular-press non-fiction literature on computer-mediated communication, 'cyberspace', and 'cyberculture'. This text-based examination draws, in the main, on the vast archive of monographs and edited collections published from the early 1990s onwards. The proliferation of published book-length accounts of the social aspects of cyberspace and cyberculture constitutes a substantial and valuable resource for tracking how on-line social interaction has been conceived and debated over time.

Also of interest here is the wider appeal of the notion of 'virtual community', as well as the flipside to this: the struggle that this notion has faced in receiving recognition in writing on the Internet within the humanities and social sciences. 'Virtual community' (like 'cyberspace') has emerged as a particularly contentious term, facing sustained challenge over its suitability as an accurate encapsulation of the diverse nature of computer-mediated social interaction. Understanding both of these things—the appeal of and challenges posed by the notion of 'virtual community'—is important in order to ascertain what significance 'community' might continue to have as an idea (and ideal), and for examining the impacts and future social possibilities of teletechnologies.

The negotiation of the key parameters and the most suitable methodological approaches to the study of cyberculture and virtual community has been fundamental to the debate. Broadly speaking, since the early 1990s the dominant strands of historical and ongoing critical concern have centred on presence and absence, questions of geographical liberation, issues of embodiment and disembodiment (due to the widely held but increasingly contested belief that cyberspace constitutes a 'bodiless' space), and identity (gender, race, sexual orientation). Prior to examining these ideas, however, it is necessary to give due consideration to the notion of virtual community: how it has been defined, how it has been commonly applied, as well as what the dominant strands of historical and present critical research into this phenomenon are.

GENERATIONS OF 'VIRTUAL COMMUNITY'

Foucault, in his (1986) essay 'Of Other Spaces', writes, 'I believe that the anxiety of our era has to do fundamentally with space, no doubt a great deal more than with time' (23). If Foucault's diagnosis encapsulates the postmodern cultural landscape of the mid-1980s, then this 'spatial' anxiety has only multiplied with the emergence of 'cyberspace'. As one critic has remarked, 'Cyberspace must be one of the most contested words in contemporary culture. Wherever it appears, the term becomes the subject of speculation and controversy, as critics and proponents argue over its function and future' (Kendrick, 1996: 143), not to mention its definition.

Science-fiction writer William Gibson (1993a: 67) coined the term, famously describing it in his novel *Neuromancer* (first published in 1984) as a 'consensual hallucination', the 'nonspace of the mind'. From this source, with its somewhat nebulous formulation, definitional tributaries have multiplied and diverged.[1] The term itself no longer strictly refers to the fictional 'matrix' of *Neuromancer* and has since entered into common parlance as shorthand for the conception of the 'space within the electronic network of computers' (Nunes, 1997: 163; Vasseleu, 1997: 46; Starrs and Anderson, 1997: 148). Even in this seemingly straightforward sense critics variously frame and debate cyberspace as a conceptual space (Whittle, 1997: 9; Rheingold, 1994: 5), a metaphysical or spiritual space (Heim, 1993; Wertheim, 1999), an alphabetic space (Tofts and McKeich, 1998),[2] a semiotic space (Shank, 1993), and a cybernetic space (Nunes, 1997: 163).

And insofar as cyber*space* constitutes an object of contemporary critical study, cyber*culture* has come to stand as the name for the field or domain of the study of cyberspace and its cultural implications and is 'concerned with new forms of social interaction made possible by computer networks' (Tofts and McKeich, 1998: 15; see also, Escobar, 1996).

These forms of social interaction facilitated by computer networks are commonly bundled together under the rubric of 'computer-mediated communication', or CMC. This is a general term referring to a range of different ways in which people communicate with one another via a computer network (Hine, 2002: 157). CMC includes both synchronous forms of communication (where the parties to the interaction need to be present simultaneously, such as with Internet relay chat (IRC)) and asynchronous forms of communication (where the parties to the interaction do not need to be present simultaneously, such as with email exchanges) (Hine, 2002: 156, 161).

Whilst still preferred by a certain section of media and cultural critics as an umbrella term, in the 1980s and early 1990s, computer-mediated communication was largely (if perhaps temporarily) supplanted within popular parlance by the more evocative metaphor popularised by the technology writer Howard Rheingold. Firing the popular imagination, his term, 'virtual community', to encapsulate the interactions made possible by CMC, has been as widely used and influential as William Gibson's coining of 'cyberspace' to encapsulate the space of networked computing.

Virtual communities, according to Rheingold's influential description,

> are social aggregations that emerge from the Net when enough people carry on [. . .] public discussions long enough, with sufficient human feeling, to form webs of personal relationships in cyberspace. (Rheingold, 1994: 5)

This definition raises manifold issues, not the least of which concern the question of what constitutes 'long enough', and how one is to determine let alone measure 'sufficient human feeling'. Nevertheless, this remains an important definition. As Jan Fernback (1999: 216) observes, few theorists have confronted the task of elucidating a comprehensive [additional or alternative] definition of the phenomenon' (see White, 2005).

A key difficulty presented by the notion of virtual community, and one likely reason for the lack of a subsequent, widely accepted definition, is that virtual community stands as an umbrella term for a range of quite diverse forms of on-line social interaction. This diversity is apparent in the genealogy Rheingold constructs in tracing the emergence of virtual community. According to Rheingold (1994), virtual community has a relatively short, albeit rich, history, dating back to the 1960s and 1970s and, most directly, to the US Department of Defense-sponsored ARPANET with its discussion list 'SF-Lovers' and email list 'Human-Nets'. Following the early success of the ARPANET lists, Rheingold argues, virtual community evolved along several lines: via computer bulletin-board systems, via multi-user dungeons (MUDs) and multi-user object-oriented systems (MOOs) (see Beaubien, 1996), and via Internet relay chat systems (IRCs), such as the well-documented French Minitel *'messagerie'* service (Rheingold, 1994: 226ff; Lemos, 1996) and the LucasFilm creation 'Habitat' (Rheingold, 1994: 188ff; Morningstar and Farmer, 1994). This is a familiar and largely orthodox historiographical account of the genealogy of on-line social interaction (Kollock and Smith, 1999; Hauben and Hauben, 1997; Kitchin, 1998; McCormack, 1998; Segaller, 1998)—albeit one that neglects older communications technologies, such as the radio, telephone, telegraph, semaphore, trains, epistolary exchange, and writing in general;[3] such an account also does not mention the important influence of the US west coast counter-culture and experimental publications such as the *Whole Earth Catalogue* on the formation of virtual communities such as the California-based WELL, or Whole Earth 'Lectronic Link (see Turner, 2006).

As testament to the appeal of the community metaphor, 'virtual community' has become even more diffuse a notion as it has entered the public realm, having grown to be something of a catchall term, encompassing everything from university alumni networks (Bentley, 1999) and company Intranet systems (Telleen, 1998), to mechanisms connecting corporate bodies to their clients (Figallo, 1998; Hagel and Armstrong, 1997). As Steve Jones (1999b: 16) remarks, 'community, particularly on the Internet, is as marketable as any other commodity'.[4]

Critical thinking and writing on the social interactions enabled by networked computing has been no less responsive to the community metaphor. But this cooption and commercialisation of the notion of virtual community opens up some interesting questions for CMC researchers. For example, David Bell (2001: 101) asks, 'why is it that commentators are so keen to see *communities* (rather than market segments or cultures of compatible consumption) in cyberspace?' A part answer to Bell's question can be found in Jan Fernback's (1999: 204) observation that, 'community, and the various meanings the word evokes [. . .] has become an efficacious symbolic term for characterizing virtual social relations'.

Rheingold's definition was formulated in the early 1990s and there has been a great deal of further thinking, experience, and research in the decade or so that followed that has shifted attitudes significantly with respect to the efficacy or 'fit' of the community metaphor. One way of summarising this change over time is to see this metaphor as passing through three general stages of development: (1) an implicit acceptance of the community metaphor as an accurate description of on-line social interaction; (2) a defensive posture of justification in light of strident criticisms of the first position, where on-line social interaction is defended as *community-like*; and (3) a more critical and introspective stance, which examines the points of difference between the metaphor and the thing it describes, and the points of contact and departure which mark computer-mediated and other forms of social interaction (that is, how is on-line social interaction or virtual community both *similar to* as well as *distinct from* community as conventionally understood?). These three stages are generalisations only. By no means do they constitute a strict chronological progression, and the application of the community metaphor within CMC research varies widely from author to author. Nevertheless, it is fair to say that the first approach has by-and-large given way to the second and, increasingly, to the third. Sketching this three-stage progression is valuable insofar as it demonstrates a development and maturation over time in thinking about and articulating the characteristics of on-line social interaction.

In the first stage the community metaphor is applied largely uncritically based on the reasoning summarised in the following passage:

> The primary reason [. . .] CMC researchers like Rheingold came up with the community metaphor to originally describe online interaction forums is that it *feels* right. Subjectively, when one looks into a virtual forum, it *feels* like what one knows as a community. (Watson, 1997: 105)[5]

As Nancy Baym argues, virtual community is community if participants *imagine* themselves to be a community (Baym, 1998: 38; see also, Bailey, 1996: 37ff, and Watson, 1997).[6] David Bell (2001: 102) views this formulation suspiciously as a kind of 'will-to-community'.

The second stage in the uptake of the community metaphor within CMC research is concerned with defending on-line social interaction as 'community-like'. This defensive posture of justification is in large measure a response to the criticisms that the vagaries of the first stage attracted. It is, in other words, a reaction to what Rick Parrish (2002: 264) terms 'objections to the medium, or those [critics] that identify some aspect of the way in which communication occurs via virtual technologies that is allegedly problematic for the concept of community'. These objections tend to pursue two general criticisms: (1) an initial questioning of the conflation of *communication* with *community*, which in turn becomes (2) an argument against virtual community on the basis of what Parrish (2002) calls the 'thinness of the medium'. The apparent conflation of communication with community can be read as an extension of the strong ties that would appear to bind communication with cyberspace (that is, cyberspace as the space of computer networks, where these networks, in addition to data and information storage, are widely used for person-to-person communication). Community formation is seen as a logical extension of the communication that takes place within the 'habitable' space of cyberspace. Coyne suggests that this line of argument is potentially borne out of a particular sense in which cyberspace is described as a 'world', which he summarises as follows:

> Cyberspace is [. . .] a world in the sense in which we talk about the world of business, academia, music, or sports—in the sense of a community and its practices. [. . .] But cyberspace is a world as community in ways other than those enjoyed by business people and hobbyists (railway enthusiasts, stamp collectors, sports fans, and the like). The claim is that electronic communications facilitate creating community through the technology itself. Telephone users rarely constitute a community simply because of a common interest in telephones. The telephone is the tool of, or means to, community. (Coyne, 1995: 153)

Thus, the argument is that, as a communicative mechanism, the Internet, like the telephone, is a tool of, or means to, community (Jones, 1998).

This qualification represents the standard response to criticisms, which tend to read the interrelationship of communication and community in virtual community as a *conflation* of communication *with* community.

A more sustained criticism of virtual community flows from the perceived shortcomings inherent to the medium. This is essentially an argument about the 'commitment' or 'stake' that is said to be a fundamental aspect of all community formation. The contention is that text-based modes of communication exchange—whether synchronous or asynchronous—do not 'provide the depth and volume of interaction necessary to create a community' (Parrish, 2002: 268; Graham, 1999). The apparent ease of withdrawal from CMC (by logging off, or switching off a computer) only serves to compound these criticisms. At this second stage, the response to such criticism is an essentially defensive (or at least interrogatory) posture—as is evident in titles like 'Are MUDs Communities?' (Bromberg, 1996), and 'Why We Argue about Virtual Community' (Watson, 1997)—which positions on-line social formation as 'community-like'.

The response to criticisms of the notion of 'virtual community' by virtual communitarians and others is really a response to the limits of the community metaphor itself.[7] As Rheingold observes (somewhat defensively):

> Even if I had the foresight to see the virtual community metaphor's deep influence, I probably would not have been able to persuade my book editor to let me use a title like People Who Discuss Things Online and Form Relationships and Groups That Resemble Community in Some Important Ways But Differ in Others. That's the beauty of metaphor: it packs a lot of meaning into a small number of symbols. But that packing process doesn't include the actual entire phenomenon the metaphor describes. In particular, when a metaphor attempts to reify human relationships, to make a dynamic process into a concrete, unchanging thing, what the metaphor fails to describe can be as important as what it imparts. (Rheingold, 1999: 18)

Or, as another critic puts it, 'metaphors are never more than partial descriptions of the phenomena they seek to describe, they are always incomplete' (Forty, 2004: 84).

This increasing introspection has proven productive insofar as it has drawn attention to the difficulties with the community metaphor, and the limitations of drawing parallels between community as it is conventionally understood, and the peculiarities and particularities of on-line forms of social interaction.

The third stage in the application of the community metaphor begins to face up to the challenge of examining the points of difference between the metaphor of community and the thing it describes, as well as the points of contact and departure between computer-mediated and other forms of social interaction (see, for instance, Baym, 2002; Slater, 2002; Jankowski,

2002: 38–40). In other words, the challenge of this third stage is to articulate what is unique about virtual community.[8] This challenge is addressed by questioning the very applicability of the community metaphor itself. That is to say, the focus lies with identifying the key points where on-line social interaction (or 'virtual community') is *similar to* as well as *distinct from* community as more conventionally understood.[9] Fernback suggests that a useful starting point in undertaking such a project is to define the phenomenon negatively, by 'articulating what cybercommunity is not':

> Community in cyberspace is not a manifestation of false consciousness. Real social practices are embedded in virtual interactions. [. . .] Cybercommunity is not commensurate with physical community in every dimension except the spatial. Virtual communities have their own cultural composition. [. . .] Cybercommunity is not just a *thing*; it is also a *process*. It is defined by its inhabitants, its boundaries and meanings are renegotiated [. . .]. (Fernback, 1999: 216–217)

As Peter Kollock and Marc Smith (1999: 23) remark, 'one can find online groups that meet any reasonable definition of community, but this is not to say that online and face-to-face communities are identical. The economies of cooperation and collective action, as one example, shift significantly as one moves to online interaction.' In other words, if 'we simply impose the term community onto all social aggregations in the virtual realm, we may miss the nuances of the virtual social experience' (Fernback, 1999: 215). Rick Parrish (2002: 259ff) makes a similar point, but even more forcefully: 'Cyberspace [. . .] supports entirely new forms of social relations to which the classic discussions of community do not immediately apply'.

Whilst the above historical three-stage progression is useful to a point (especially in summarising the evolution of the community metaphor within the literature on CMC), it leaves the challenge of understanding and defining the difference between the notion of 'community' independent of 'virtual community'. Such is the scope and importance of this issue that it is addressed in detail in its own right in the following chapter. It also leaves the challenge of examining the key interrelated themes that are central to most (if not all) of the debates about virtual community and computer-mediated communication. These are: presence, geographical liberation, dis/embodiment, and identity. Each will be explored in turn below.

KEY THEMES IN VIRTUAL COMMUNITY DEBATES

Presence

A defining feature of cyberspace, and of 'telematic' society as a whole, is that 'communication, identity and presence are defined by absence' (Tofts and McKeich, 1998; Wood, 1998). Simply put, our engagement with cyber-space would seem to operate at a distance. Central to this understanding of absence and distance are the interrelated notions of 'presence' and 'telepresence': in general terms, this is the *effect* of being in a particular place while actually being somewhere quite different. In the literature on the construction of virtual environments, these two terms—'presence' and 'telepresence'—have historically been applied in contradictory ways, with telepresence used both interchangeably with and in opposition to the term presence, and presence sometimes considered as an overarching term that is inclusive of telepresence.

The word 'telepresence' was coined by Marvin Minsky in 1980 'in refer-ence to teleoperation systems for remote manipulation of physical objects' (Steuer, 1992: 76)—such as we see with the use of remote controlled 'drone' aircraft operated in Afghanistan from remote locations. After Minsky, early applications of the term 'reserv[e] telepresence only for cases involving teleoperation', and use presence 'to refer to the generic perception of being in an artificial or remote environment' (Steuer, 1992: 76). Thomas Sheridan (1992: 120) prefers to employ 'presence' as an umbrella term which incor-porates 'telepresence' ('feeling like you are "there" at the remote site of operation') and 'virtual presence' ('feeling like you are present in the envi-ronment generated by a computer'). Jonathan Steuer, meanwhile, proposes that presence and telepresence should in fact be differentiated. The former refers to 'the sense of being in an environment':

> Presence can be thought of as the experience of one's physical environ-ment; it refers not to one's surroundings as they exist in the physical world, but to the perception of those surroundings as mediated by au-tomatic and controlled mental processes. (Steuer, 1992: 75)

The latter, telepresence, is, 'the extent to which one feels present in the mediated environment, rather than in the immediate physical environment' and is defined as 'the experience of presence in an environment by means of a communication medium':

> When perception is mediated by a communication technology, one is forced to perceive *two* separate environments simultaneously: the physical environment in which one is actually present and the environ-ment presented via the medium. The term *telepresence* can be used to describe the precedence of the latter experience in favor of the former. (Steuer, 1992: 75–76)

For Steuer, then, presence refers to an apparently 'unmediated' experi-ence of one's surroundings, while telepresence refers to a 'mediated' experi-ence of the same. It is this formulation of Steuer's—a 'mistaken dichotomy between, on the one hand, "real", "natural" presence and, on the other hand, "mediated" telepresence' (Milne, 2010: 166)—that has drawn criticism. The difficulty with this formulation, as Giuseppe Mantovani and Giuseppe Riva (1998) note, is that it fails to acknowledge that presence is 'always mediated' and that it is culturally constructed.

This is where William J. Mitchell's (1999a) notion of an 'economy of presence' is useful. Mitchell's deceptively simple proposition is that *all* communications technologies circulate within an 'economy of presence' where each communication option (face-to-face encounter, telecommunication, telepresence) should be assessed according to need.[10] Which option is selected will reflect 'the balances and combinations of interaction modes that turn out to work best for particular people, at particular times and places, facing their own specific circumstances within the new economy of presence' (Mitchell, 1999a: 144).[11] As Tofts (2000a) observes, Mitchell's 'economy of presence' is 'a decisive and foundational notion for reflecting on the determination within our daily social reality of what forms of communication are relevant or appropriate'.

But semantic debate over the notion of presence does not stop here. Matthew Lombard and Theresa Ditton's (1997) taxonomic study delineates at least 'six interrelated but distinct conceptualizations of presence'. These are, in brief:

1. 'presence as social richness' ('the extent to which a medium is perceived as sociable, warm, sensitive, personal or intimate when it is used to interact with other people'; 'presence as social richness is related to two important concepts [. . .]: intimacy and immediacy');
2. 'presence as realism' ('the degree to which a medium can produce seemingly accurate representations of objects, events, and people');
3. 'presence as transportation' (the idea of being transported to another realm, which can be broken down into three subtypes: 'you are there', 'it is here', 'we are together');
4. 'presence as immersion' (which 'emphasizes the idea of perceptual and psychological immersion');
5. 'presence as social actor within medium' (where 'users' perceptions and the resulting psychological processes lead them to illogically overlook the mediated or even artificial nature of an entity within a medium and attempt to interact with it'); and
6. 'presence as medium as social actor' (which 'involves social responses of media users not to entities [people or computer characters] within a medium, but to cues provided by the medium itself', and which leads users 'to treat the medium as a social entity'). (Lombard and Ditton, 1997)

What Lombard and Ditton discover from their survey is that all six definitions, while quite different, share a central idea: 'the perceptual illusion of nonmediation'. This 'occurs when a person fails to perceive or acknowledge the existence of a medium in his/her communication environment and responds as he/she would if the medium were not there' (1997). It is an important understanding for grasping the way that computer-mediated communication exchange operates and how the absence and distance that characterises this exchange can be felt, albeit by illusion, to be experientially 'overcome'. This illusory experience of nonmediation through presence (especially in the sense of 'presence as transportation', or the effect of being in a particular place while actually being somewhere quite different) contributes in key ways to the second theme: the notion that cyberspace affords true geographical liberation.

Geographical Liberation

'New technologies,' David Morley (2003: 439) observes, 'have been widely credited with producing a range of transformative effects on the way we live.' One of these effects is that 'new communications technologies have been trumpeted as heralding the ultimate "death" of geography' (439). One sense in which this view is promulgated in the literature on cyberspace and CMC can be linked to the conception of cyberspace as 'a "space" that has no coordinates in actual space' (Tofts and McKeich, 1998: 15), a space 'parallel to actual space' (Benedikt, 1992b). But the main sense in which communications technologies—particularly computer-mediated communications technologies—are commonly held to transcend geography is tied to a consideration of cyberspace as a text-based medium (or what Rick Parrish (2002: 260) somewhat awkwardly calls a 'post-geographical context-medium'). This basic mode of textual exchange poses particular problems for those attempting to 'apply traditional communitarian thought to the virtual arena' (Parrish, 2002: 261). The difficulty in 'accepting the word "community" as a descriptor for CMC phenomena,' Nessim Watson (1997: 120) suggests, is that, 'in CMC, physical space has been replaced by a technology, a medium of communication'. For Parrish at least, what is needed in response is to redefine how we think about space and place:

> Cyberspace is not a geographic entity, and until "our sense of place becomes post-geographical"—that is, until we begin to comprehend how common understandings of everyday concepts are not necessarily adequate to a universe which is constructed of fundamentally different material—we will continue to naively assume that virtual community must mimic traditional community to achieve those ends which we agree upon as the legitimate focus of the concept. (Parrish, 2002: 260)

These observations intersect with wider philosophical meditations on the problematics of community, where the very grounds on which community is possible have been questioned (see the following chapter).

Within CMC research, the tension between traditional conceptions of community and 'virtual community' has led to the formulation of definitions of community that are said to be 'both recognizable to most traditional thinkers and permissive of interactive electronic media' (Parrish, 2002: 260). What such formulations amount to is a downplaying in significance of geography/space/place within formulations of community in favour of an emphasis on other constitutive features of community-formation, such as interpersonal interaction, common ties and shared interests (as well as other factors like membership and social control and policing). There is also an emphasis on the means to express these ties and communicate these interests (Parrish, 2002: 260–264; Watson, 1997: 120–126ff; Jones, 1997).

The apparent drive to dissociate on-line social interaction and 'virtual community' from geography has implications for the individual as much as for the collective. These include a desire, at least in early literature on cyberspace, to transcend the body and to experiment with identity—both of which form the focus of the third theme.

Dis/Embodiment and Identity

Questions concerning dis/embodiment and the construction of identity have been central to CMC research. The first of these concerns, in its crudest and most extreme formulation, is fuelled by the separatist vision of cyberspace as an abstract realm with no coordinates in actual space, and the fact that social interaction is primarily text-based (and emotion is expressed and experienced via 'emoticons'). This leads some techno-boosterists to conceive of cyberspace as an 'ecstatic' space,[12] a celebration of the medium of cyberspace as a 'disembodied' realm that is free from the constraints of physics and the messy contingencies of the flesh and in which there is a 'sense of disembodied communion' (Sconce, 2000: 21).[13] Rheingold (1993: 58) enthusiastically declares that in cyberspace, 'we do everything people do when they get together, but we do it with words on computer screens, leaving our bodies behind' (see also, Weimann 2000: 334). Among the more contentious extrapolations of this idea is Neil Spiller's (1998a: 50) assertion that 'virtually inhabiting cyberspace [. . .] creates a transhuman condition where it is possible to escape our frail, mushy, non-responsive, decaying, wet bodies for certain periods of time.'[14] Spiller dubs this process 'visceral escapology', and he derisively labels 'flesh chauvinists' and 'flesh Luddites' (1998a: 137, 139) those who quibble about the morcellation and abandonment of the body. Suffice to say, such provocative and largely unsupported (and insupportable) assertions have by-and-large dissolved, or been discredited. As Tofts (2004b: 32) puts it, 'interacting with computers is a completely embodied experience' and that, as others note, 'even in cyberspace, people are people' (Schiano and White, 1998: 359; Benford, et al, 1995).[15]

The motivations behind, and darker implications of, this desire for disembodiment are examined by N. Katherine Hayles. 'Why do we want to leave the body behind?', Hayles (1996: 2) asks. Beyond a vision for disembodiment being 'nurtured by a cultural tradition that has long dreamed of mind as separate from body', the answer, Hayles suggests, is in large part an 'objectionable . . . escapist fantasy' (3). Scratch the surface of this desire for disembodiment, she writes,

> and you will find scarcely concealed anxiety about our continued existence on a planet despoiled by environmental poisons and decimated by AIDS. Ironically, such fantasies may be complicit in furthering the very anxieties that engender them. If we can live in computers, why worry about air pollution or protein-based viruses? (Hayles, 1996: 3)

This fantasy also has implications for gender politics:

> Inscribing the female power of reproduction into a technological scenario of (male) parthenogenesis, it identifies subjectivity with the rational mind that has traditionally been encoded masculine, leaving behind the materiality of the body that has been identified with the feminine. (1996: 3)

Referring back to historical studies in physiognomy that sought to show that 'women and blacks had smaller heads, and therefore supposedly less capable brains, than white men', Hayles suggests that 'similar metaphoric mappings underlie the erasure of the body from cyberspace':

> The dualities line up as follows: mind is superior to body; silicon technology is superior to protein organism; man is superior to woman. Therefore replace the body of woman with a computer that can serve as a fitting receptacle for the (male) mind. The privileged terms (mind, computer, male) are linked together in mutually reinforcing connections that seem to make it possible to erase or leave behind the stigmatized terms (body, organism, female). (1996: 4)

Hayles' telling conclusion is that the positioning (or vision) of cyberspace as a disembodied realm is not intrinsic to the medium. Nor is it innocent. Rather, it 'happens in two ways simultaneously, through technological interventions and discursive practices' (4).

A somewhat more subtle (but no less problematic) variation on the disembodiment theme is the claim that cyberspace enables identity to be constructed in ways categorically different from face-to-face interaction. In cyberspace, Rheingold (1993: 58) writes, 'our identities commingle and interact electronically, independent of local time or location'. 'An early promise of online interaction,' Kollock and Smith (1999: 10) recollect, 'was that it would render irrelevant such markers as race, gender, status, and age'—as Weimann (2000: 335) writes, 'In the virtual world, one is neither expected nor required to keep real-world attributes, such as age, appearance, profession or gender'. Consequently, the virtual world becomes a place of experimentation and exploration'. 'In these narratives of freedom', Esther Milne (2007: 2) argues, 'the subject is fragmented, multiple, self-determining, and therefore in a "position", so to speak, to refuse to occupy the power structures of hegemonic subject positions'. Moreover, in an examination of race in Usenet, Byron Burkhalter observes:

> In face-to-face interaction an individual's physical characteristics, from skin color to vocal patterns, help convey racial identity. Lacking such physical cues on computer networks, one might predict that discrimination on the basis of race, age, gender, sexuality, class, status, and group membership would disappear. Indeed, some participants use the lack of physical cues to claim any identity they want. (Burkhalter, 1999: 63)

The 'absence of these indicators' in CMC exchange is said to allow 'online subjectivity to be playful, performative, flexible and decentred' (Milne, 2007: 3; see also, Reid, 1995; Turkle, 1995; Parrish, 2002: esp. 273–279; Stone, 1995; Spiller, 1998a: 69). According to this view, identity is 'mutable' (Jones, 1995: 5): there is no essential 'rational, autonomous, centred and stable core' to identity (Poster, 1995: 80). Other writers argue, in the words of Milne (2007: 3), that 'computer-media strips bodies clean of the marks of culture, revealing the "real" self'.

The claims made in respect to on-line identity are also said to be true of more general online behaviour, where cyberspace is conceived as a kind of Bahktinian 'carnivalesque' space, or 'liminal' space: 'the Net [is] a unique place that allows for the transgression of cultural rules, the breaking of taboos, the freedom to express what you need to, when you need to [. . . and where] all codes of conduct, all rules of behaviour are abandoned' (Argyle, 1996: 137).

But as Milne (2007: 3) observes, 'a substantial body of research has questioned the technological production of a decontextualised, incorporeal, genderless, raceless and ageless subject'.[16] For instance, Burkhalter (1999) argues that race is just as 'relevant' in on-line social interaction. Similarly, Jodi O'Brien (1999) argues that gender is such a central feature for organising interpersonal relations that most users of CMC are at pains to reproduce gender in on-line interaction.[17] Kollock and Smith (1999: 9) echo these views: 'traditional status hierarchies and inequalities are reproduced in online interaction and perhaps are even magnified'. And in contrast to writers who celebrate the supposed liberating potential and 'decentred, fragmented selves of CMC' (Milne, 2007: 3), Beth Kolko and Elizabeth Reid (1998: 213) argue that 'fragmented projections of the self can become fixed and invariable, and can preclude flexible social interaction'. Moreover, in contrast to the celebration of the Internet as a socially transgressive 'free space', where all rules of behaviour are abandoned, Reid elsewhere argues that the prevalence of 'flaming', and an emphasis on humiliation, castigation, retribution, and ostracism in many MUDs, 'shows a return to the medieval' in terms of technologies of social control, discipline, and punishment (Reid, 1999: 118, 107–133). Further research also questions the strict dichotomy sometimes assumed to exist between 'on-line' and 'off-line' identity. As Nancy Baym (1998: 55) notes, 'the reality seems to be that many, probably most, social users of CMC create online selves consistent with their off-line identities'.[18]

COINCIDENCE AND DIFFERENCE: FROM 'VIRTUAL COMMUNITY' TO 'COMMUNITY'?

Throughout the 1980s and 1990s, the notion of virtual community was a popular but contested metaphor for encapsulating what is involved in and unique about CMC exchange. The appeal of and to the community metaphor in the literature on CMC and the Internet serves to emphasise E. H. Gombrich's point (included as an epigraph to this chapter) that 'the familiar will always remain the likely starting point for the rendering of the unfamiliar' (1972: 72). In other words, what we know shapes what we hope to know and understand. In this particular case, and for better or worse, the notion of community has shaped in profound ways how CMC and a diverse range of other forms of on-line interaction have been conceived and debated. As David Bell and Gill Valentine note, 'the term community is not only descriptive, but also normative and ideological: it carries a lot of baggage with it' (quoted in Bell, 2001: 93).

Moreover, the popularity of the community metaphor, particularly in the business press, has been such that it risks losing what efficacy it may once have had as a descriptive term.

One effect of this popularity and proliferation in usage has been a more careful critical consideration over time of the appropriateness or 'fit' (or otherwise) of the community metaphor for understanding the nature of CMC exchange. Thus, from its earliest usage to the present (as discussed earlier in this chapter), application of the community metaphor in virtual discourse has passed through several key phases, from an initial and almost unquestioning embrace to a more circumspect and defined usage where it is considered to be similar in some ways but different in other key respects from community as more generally and traditionally understood. On the one hand, points of similarity or convergence between virtual community and community as more commonly understood are most evident when 'community' is taken to mean and involve social interaction, information exchange, common ties, and shared interests. On the other hand, virtual community is said to be distinct from broader conceptions of community through the following factors: its reliance on predominantly textual cues for expressing meaning and affective expression; the creation of a sense of communicative presence achieved at a distance; and claims for the development of a unique set of discursive codes and practises.

At very least, what these points of similarity and difference point towards is maturation in how virtual community and computer-mediated communication are imagined and debated.

Nevertheless, since the mid-1990s, research on virtual community and CMC appears to have shifted from questions of legitimation to more detailed critical considerations of issues of embodiment and disembodiment, presence and absence, identity (gender, race, sexuality, etc.), and questions of commitment or stake in on-line interactions. In this work, many of the early and enthusiastically 'boosterist' (and at times outrageous) claims regarding disembodiment, identity, and geographical liberation have undergone careful, sustained, and detailed critique. The result of this expanding body of critical literature is greater detail and subtlety of understanding in research knowledge on computer-mediated social interaction. There is growing recognition that they will 'continue to gain social acceptance and internal solidarity, cutting across existing communities as well as coinciding with them' (Rooksby, 2002: 135; see also Etzioni and Etzioni, 1999). This has important social and political implications:

> We can take steps to make them coincide, and it can be socially and politically worthwhile to make the effort to do so, but the social effects of non-coincidence may well be to disrupt any sense that one belongs to and must work with and within any community at all. (Rooksby, 2002: 135)

However, greater coincidence carries its own risks. Claims for the uniqueness of virtual community, as Gordon Graham (1999: 149) points out, 'Actually *depend* upon its limitations as a form of communication'. These are limitations which are imposed by the medium and by the constraints of present technologies. 'It is these limits,' Graham argues, 'which allow participants to disguise the personal properties—gender, skin colour and so on—which would hinder freedom of exchange.' (149) His conclusion is that, 'the more these limitations are overcome, the more Internet communities look like ordinary communities with all their disadvantages—the more, in short, virtual communities come to look like the normal thing' (149–150).

In addition to this, some further criticisms of virtual community are that it only allows for weak social ties (Kavanaugh, et al, 2005; Haythornthwaite, 2002), that it lacks social stake and commitment by permitting ease of withdrawal, and that it is therefore socially and politically impoverished (or anaemic) in that it evades—while never fully escaping—the complexities of what Doreen Massey refers to as the 'question of our throwntogetherness', the 'unavoidable challenge of negotiating a here-and-now' of (involuntary) social interaction (Massey, 2005: 151, 140; see also, Robins, 1999; Robins and Webster, 1999). However, the possibility that Graham proposes above impacts significantly on such criticisms.

Admittedly, to conceive of virtual community as one form of community among others by no means nullifies such criticisms. But it does call for a qualification—perhaps even a recasting—of them. This is particularly so given the extent to which some of the more outlandish or questionable assertions regarding virtual community and teletechnologies are continuing to be challenged and qualified.

This observation finds a further degree of support in Baym's (1998: 55) finding that in most cases participants in virtual communities do not in fact tend to disguise gender, race, and other social cues, and that one's on-line persona (or personae) is generally consistent with their 'off-line' presentation of self. In other words, the apparent radicalness or transgressive appeal of CMC exchange is arguably—and perhaps, *significantly*—overstated. Indeed, the extent of on-/off-line consistency would seem to suggest that several markers of difference—such as anonymity and experimentation—which are often assumed to be key to distinguishing 'virtual community' from 'community' begin to evaporate (or at least are of diminishing significance). Such an elision of difference suggests that Espen Aarseth (1997: 146) has a point in questioning the usefulness of the term 'virtual' in the phrase 'virtual community'.

However, to continue to credit and maintain what can legitimately be considered unique about on-line social aggregations suggests there is value in retaining some kind of linguistic or semantic marker of differentiation (such as the word 'virtual'), while reformulating virtual community's relationship with wider conceptions of community. This is particularly important since casting virtual community as a valid form of community opens it up to the same philosophical challenges that face 'community' in general. The more this occurs, the more urgent becomes the need to reformulate the very concept of 'community' in non-restrictive terms. It also raises the question of how the increasingly extensive interactions between forms of community and layers of technological mediation can be productively rethought (see the final chapter).

The aforementioned issues also lead inexorably towards 'arguments about community as a whole' (Bell, 2001: 93). As noted earlier, any consideration of virtual community needs to be set in the broader context of 'a critical approach to the concept of "community" in late modernity' (James Slevin quoted in Bell, 2001: 93). This is particularly important given, as noted earlier, that 'the term community is not only descriptive, but also normative and ideological: it carries a lot of baggage with it' (James Slevin quoted in Bell, 2001: 93). Examining broader debates about the concept of community, and all its associated 'baggage', is the task of the next chapter.

2 The Problem of Community

'... but isn't community outside intelligibility? ...'
—Maurice Blanchot (1988: 1)

In the course of the previous discussion on the notion of virtual community, the idea of community, for those writing on teletechnologies and especially computer-mediated communication (CMC), emerged as both an evocative and contentious notion. The aim of this chapter is to understand how the concept of community is developed and debated in texts and contexts that are outside of, and in some cases predate, the literature on the virtual and CMC.

An examination of community is valuable for its ability to broaden the focus of the previous chapter by situating the arguments over virtual community within a broader history of intellectual thought and debate regarding the notion of community. Thus, the discussion here aims to further test the understanding, introduced in the previous chapter, of 'community' as a concept that is appealing yet contentious, and to shed further light on the notion of 'virtual community' through examining the efficacy (or otherwise) of 'community' as a term (or metaphor) for encapsulating this particular nature of on-line social interaction.

But how to do so? Clearly it is neither possible nor desirable to attempt a comprehensive coverage here of the vast array of different (and often conflicting) theoretical considerations of a subject as broad as 'community' (see Delanty, 2003), to give a full account of historical and contemporary formulations of community that may prove most productive for thinking about the social aggregations that form through computer-mediated social relations (see Willson, 2006).

Rather, the path that is taken here is to adopt a cross-sectional approach by examining three distinct historical points within the twentieth century where community emerges as an important yet contested notion. More precisely, within each of these three sections the focus is on three sets of perspectives on community that are quite different from one another and which are instructive for how they take up the issue of community. Material is drawn from the disciplines of sociology and continental (especially poststructuralist) philosophy, as well as, to a lesser extent, from political science.

A cross-sectional approach is useful insofar as it provides a range of perspectives on the notion of community, which can shed further light on the paradoxical nature of community as a concept that is appealing yet contentious. These perspectives each bring a different inflection to, and understanding of, the idea of community. Furthermore, these perspectives, while quite different from one another, when combined provide a composite picture of community.

The perspectives selected, and the rationale for these selections, is provided in the introduction to the three subsections below. A key factor in their selection has been how each relates to and sheds further light on the notion of 'virtual community' as developed thus far. Therefore, as each perspective on community is examined, implications for virtual community will be considered at the same time. Specifically, the discussion focuses on themes from these writers (and relevant critique of and commentary on them) that reflect and contribute to contemporary understanding of the concerns of virtual community.

A number of preliminary and general observations about the notion of 'community' serve to introduce this three-part examination of the concept of community. This material both elucidates the particular approach that is taken in this chapter and it provides important contextual background to the notion of community.

CONCEPTUALISING COMMUNITY

Community is a term that most people have a general familiarity with. As a concept, it tends to be intuitively grasped, but proves far more elusive to define (see König, 1968: 14–30; Hughes, Michell, Ramson, 1992: 221).

Community is also a particularly evocative and emotive idea that has captured the imagination of specialist and non-specialist alike (Thorns, 1976: 15). It is 'one of those "motherhood" words which, like democracy, welfare and participation tends to be accepted as indubitably a good thing' (Bryson and Mowbray, 1981: 256). It is a word that 'feels good' (Bauman, 2001: 1). Raymond Williams (1976: 66) suggests, 'what is most important, perhaps, is that unlike all other terms of social organization (state, nation, society, etc.) it seems never to be used unfavourably and never to be given any *positive* opposing or distinguishing term'. For at least one critic, this is community's 'blind utopianism' (Van Den Abbeele, 1991: ix).

Yet, community is also generally understood as something lost and for which we are nostalgically yearning. According to Zygmunt Bauman (2001: 3), '"community" is nowadays another name for paradise lost—but one to which we dearly hope to return, and so we feverishly seek the roads that may bring us there'. Similarly, Robert Nisbet (1953: 47) writes that, as 'the quest for community', this cycle of perceived loss of and yearning for community is 'timeless and universal'.[1] If this is the case,

> [and] if we are to understand the conditions that lie behind the quest for community in our society, we must look not merely to contemporary

social and psychological dislocations but to the historical sequences of change which led up to them. (Nisbet, 1953: 77)

Two of these 'historical sequences of change' are sketched below.

The first key historical sequence underpinning and feeding the contemporary 'quest for community' is widely taken to be the strong separation in the West of 'the social thought of the nineteenth century from that of the preceding age, the Age of Reason' (Nisbet, 1967: 47; see also, Gusfield, 1975: 1–20; Todorov, 1996). It is a shift characterised by the nineteenth century 'rediscovery' of community as a 'revolt' against Enlightenment individualism (Nisbet, 1967: 7, 47). To trace the quest for community through this shift is to reveal (and unravel) a complex lineage, taking in the work of such diverse thinkers as Hobbes, Burke, Rousseau, Carlyle, Comte, de Tocqueville, Le Play, Weber, Marx, Durkheim, and Simmel, to name a few (see Donini and Novack, 1982; Nisbet, 1953, 1967). To do justice to such an endeavour would lead this discussion too far afield. For the present purposes, therefore, the work of Nisbet (1953, 1967, 1969), a key commentator on the idea of community within the history of social theory, reviews and summarises many of these issues.

Nisbet (1967: ix) writes that, 'the most fundamental ideological conflicts of the past century and a half [to two centuries] have been between, on the one hand, values *of community* [. . .] and, on the other hand, *individualism,* [. . .]'. This conflict appears almost cyclical in its shifts from the one to the other—the individual to the communal and back again—from the eighteenth to the nineteenth century and into the twenty-first century.

Nisbet suggests that individualism re-emerges in the early twentieth century in Modernism (in its non-architectural sense), albeit as disenchantment, in the light of the devastation wrought by the two world wars. As one critic puts it, 'After World War I [. . .] the autonomous freedom which attended this individualism suddenly seemed dangerous. More than that, the loneliness which attended this individualism finally seemed more than we could bear' (Rouner, 1991: 2). In response to this anxiety and disenchantment, 'the quest for community become[s] the dominant social tendency of the twentieth century' (Nisbet, 1953: 45)—including community as manifest in totalitarianism and, specifically, in its most confronting configuration: the 'community' of German National Socialism as it culminated in Nazism (33–34).

For Nisbet (writing in 1953), the driving force behind this persistent return to community, particularly in the post-WWII period, is not nostalgia but rather 'the failure of our present democratic and industrial scene to create new contexts and new associations and moral cohesion' (1953: 73). Nisbet is at pains to stress that technology is not to blame for this failure (74). The key historical reason, in his view, is the rise and structure of the Western political state (47) and the concomitant isolation of the individual citizen from any 'sense of meaningful proximity to the major ends and purposes of his culture' (73), and where the 'primary social relationships are increasingly functionless, almost irrelevant with respect to these ends' (52).[2]

A second, key historical sequence of change occurs in that period of the latter part of the twentieth century from the mid-1980s onwards—the period which Fredric Jameson (1991) terms 'late capitalism'—when a call for a return to community re-emerges during this time (Van Den Abbeele, 1991: x). Renewed interest during this period in community *per se* is due in large part to the fact that, within postmodernism and its associated 'ambiance of vertiginous transformation and individualization', it is a necessity that 'at least a minimal claim of transpersonal relevance must be made if there is to be *any* politics at all' (Van Den Abbeele, 1991: x–xi).

Yet, contemporary consideration of the question of community argues that this cyclical turn away from and return to community becomes truly cyclical in a fixed sense; the circle is not a forward-moving spiral, let's say, and the concept of community is not progressed in any meaningful or constructive way (see Miami Theory Collective, 1991).

It is argued that the reason for this cyclical entrapment is that there is a 'demonstrable paucity of ways to think community' within the European and Anglo-American tradition (Van Den Abbeele, 1991: ix). Both the conceptual paucity of community and its 'ecumenical appeal' are revealed in Georges Van Den Abbeele's overview of the two popular etymologies of the term. On the one hand, he argues, there is the 'more philologically valid formation of the word from *com + munis* (that is, with the sense of being bound, obligated, or indebted together)' (1991: xi), while, on the other hand, there is the 'more folk-etymological combination of *com + unus* (or what is together as one)' (xi). The point here is that 'the stakes involved in choosing between a community that is mutual indebtedness and a community that is absorption into oneness are more than just philological' (xi). The import of this choice, and what is hidden behind this choice, is borne out by the following 'coincidence':

> The [two] rival etymologies point to the two classic ways the West has tried to theorize community, between the organicist notion of the 'body politic' most colloquially linked with the name of Hobbes *and* the idea of the social contract popularized by Locke and the Enlightenment *philosophes*. (Van Den Abbeele, 1991: xi)

Van Den Abbeele argues that the difficulty with these two politico-philosophical conceptions of community is that each, in its own way, fails to come to terms with how 'freely engaged subjectivities are constituted'; it fails to account for 'the difference between singular subjectivities which is part of what they share by being in common' (xii). His argument is that alternatives to community, such as atomism and totalitarianism, 'have each proceeded to an aggressive reduction and elimination of social difference, which in turn has fuelled the contemporary sense of the loss of community' (xii).[3]

This problem is summarised neatly by Linnell Secomb (2003b: 9) when she writes that, on the one hand, 'the formation and perpetuation of in-common community requires the estrangement of those who threaten its commonality.' In other words, existing (Anglo-American and European) conceptions of community fail at 'remaining open to the other' (Lucy, 2001: 146). Yet, on the other hand, 'community is nonetheless indispensable. Human existence is social existence [. . .]. Human being is a being-together-in-community' (Secomb, 2003b: 9).[4]

Thus, the *paradox* of community—the aforementioned cycle of loss and yearning for community—in these postmodern times becomes the *problem* of community.

The questions this problem poses are as follows. How to (re)negotiate this theoretical impasse? How to formulate a model of community that accommodates the 'other'? How to think afresh a model of in-common community that welcomes the 'other' which threatens its very commonality?

It is these issues which lie behind Blanchot's (1988: 1) questioning of the very intelligibility of 'community'.

In an academic context, the question of community and the challenge of making community 'intelligible' generates multiple perspectives and spans a variety of disciplines, including sociology, philosophy, political science, and urban planning, to name a few. These perspectives may overlap, or they may remain separate. Moreover, some of these perspectives, and the factors contributing to their formulation, come into sharper focus against specific backgrounds and at specific points in time.

Having provided this preliminary overview of the concept of community, discussion now turns to the three different sets of perspectives on community.

CROSS-SECTIONS OF CONCEPTS OF COMMUNITY

The following perspectives on community are drawn from three different and evenly spaced moments in recent history: early, mid, and late twentieth century. In the first period, the ideas of German sociologists Ferdinand Tönnies and his contemporary, Herman Schmalenbach, are examined. In the second period, two key essays from the American sociological literature of the 1950s are examined. In the third period, the work of the French poststructuralist philosophers, Jean-Luc Nancy and Jacques Derrida, is examined.

Unravelling Totality: Gemeinschaft/Gesellschaft and the Bund

The works of Ferdinand Tönnies, a major figure in twentieth century sociological theory, and of Herman Schmalenbach are important reference points in the development of ideas of community within classical social theory. Tönnies's dualist conception of community and society has proven immensely influential in broader social theory and more recent thinking on social structure. Schmalenbach was a contemporary of Tönnies and, while lesser known, has been selected here for the way in which he responds to Tönnies's conception of community, both refining it and extending it further through the insertion of a third, middle term bridging those of community and society (Hetherington, 1998). Most importantly, and as shall become evident, his theories have direct relevance to thinking about community in the context of teletechnologies (Bell, 2001: 107).

The intellectual heritage and the broad historical context informing the development of both thinkers' ideas are rich and complex.[5] Their period of works (the late nineteenth and early twentieth centuries) was characterised by large-scale industrialisation and urban expansion. With these changes came the pressing question of how to account for the increasing disjunction between small-scale social and communal existence and large-scale, often socially isolated, urban existence. This is the context in which Tönnies was writing, and in response to which he developed his ideas of *Gemeinschaft* and *Gesellschaft*.

In his conception of *Gemeinschaft* and *Gesellschaft*, Tönnies (1963) created what has become one of the most distinct and enduring modern formulations of community and of that which takes us away from community. *Gemeinschaft* translates readily enough as 'community' (1963: 33). *Gesellschaft* is generally translated as 'society' (33), but the term is in fact more elusive (and allusive). According to Nisbet (1967: 74), *Gesellschaft* is 'characterized by a high degree of individualism, impersonality, contractualism, and proceeding from volition or sheer interest rather than from the complex of affective states, habits, and traditions that underlies *Gemeinschaft*'. But even in this more nuanced interpretation of *Gesellschaft*, the pejorative inflection given it by Tönnies *vis-à-vis Gemeinschaft* is apparent:

> [Gemeinschaft] is the lasting and genuine form of living together. In contrast to Gemeinschaft, Gesellschaft is transitory and superficial. (Tönnies, 1963: 35)

This is reiterated in the suggestion that the city (or metropolis) is seen as a paradigmatic instance of *Gesellschaft*, albeit with (seemingly) dire consequences for *Gemeinschaft*:

> But as the town lives on with the city, elements of life in Gemeinschaft, as the only real form of life, persist within the Gesellschaft, although lingering and decaying. (1963: 227)

In a significant later move, Tönnies extends *Gesellschaft* from the city-metropolis to capital and the formation of the nation state (228 & 234).

In making this extension, Tönnies suggests that *Gesellschaft*, in its more expansive sense, works against *Gemeinschaft*, but speculates (presciently as it turns out) that the 'success of such attempts is highly improbable' and that (interest in) community will rebound.

The great difficulty with Tönnies's model, at least as it has filtered down to current sociological usage, is its reduction to simplistic conceptions of its two central terms: community and/vs. society. It is, of course, a much more complex structure than this reduction to simple binaries allows.[6] Suffice it to summarise and stress the aforementioned points as key in the *Gemein-schaft/Gesellschaft* model. First, the distinction between these two ideas clearly appeals to the prevailing (or at least enduring) sentiment of community as simultaneously 'lingering and decaying' (if not lost), but at the same time it emphasises community as a concept that is resilient and which will likely rebound. Secondly, the alignment of *Gesellschaft* with capital and the emergence of the nation state is a move that (at least for Nisbet and, as we saw above, for Van Den Abbeele also) is crucial to understanding the (re)turn to community in the mid-twentieth century.

One little considered approach to rethinking community in light of the above can be found in the work of Tönnies's contemporary, the German sociologist Herman Schmalenbach and his work on the concept of the 'Bund'.

For English-language readers, Schmalenbach's understanding of the Bund is primarily remembered—if it is remembered at all—as a critique of Tönnies's *Gemeinschaft/Gesellschaft* model, where it is presented as a 'third term' to supplement and correct Tönnies's famous dualism: that is, *Gemeinschaft*, *Gesellschaft* and the Bund (see Schmalenbach, 1961). Through the addition of the Bund, this trichotomy, rather than dichotomy, offers 'a less rigid and more cyclical view of social change than the unilinear one offered by Tönnies' (Hetherington, 1998: 89).

The concept of the Bund was first developed by Schmalenbach in his (1922) essay, 'Die Soziologische Kategorie des Bundes'. The first English translation of Schmalenbach's essay appeared in a widely circulated anthology in 1961 as a truncated translation which emphasised Schmalenbach's critique of Tönnies. An English translation of the original essay in its complete form did not appear until 1977 (see Schmalenbach, 1977). This second, full translation contained much more than just a critique of Tönnies.[7] For the purposes of the present discussion, however, emphasis will be placed on Schmalenbach's critique of Tönnies's model.

The term 'Bund' has a 'long and convoluted history' (it comes from the Indo-Germanic verb *bhend*, meaning 'to bind' or 'to tie') but has, over time:

> come to be associated with a range of formal and informal types of organisation, all of which in some way imply a sense of looseness in terms of elective membership while at the same time suggesting the strength both of loose organisational forms and of the social bonds they create.
> (Hetherington, 1998: 84)

The question pursued here is: How does the concept of the Bund address the aforementioned 'problem' of community, the double-bind of community being indispensable yet simultaneously estranging the 'Other' who threatens its commonality? Moreover, what relevance does the concept of the Bund have for understanding *virtual* community?

Useful responses to both these questions can be found in the work of Kevin Hetherington, who has revived and consistently championed the contemporary significance of Schmalenbach's concept of the Bund.

One of the primary reasons the Bund is considered by Hetherington to have ongoing relevance is that it is 'based [. . .] in feeling and emotion rather than in the more instrumental practices usually associated with organisations' (1998: 84). This foundation emerges from 'Schmalenbach's aim [. . .] to differentiate between traditional and affective forms of conduct, with the Bund as a form of sociation associated with affective social action' (90). It is in this understanding of affective sociation that we find the root of Schmalenbach's critique of Tönnies and what he sees as the latter's 'romantic and confused longing for community' (89). As Hetherington explains, what this longing seeks is:

> the elective, affective-emotional solidarity of the Bund. [. . .] Gemeinschaft is based in the unconscious [. . .] belonging to a group [. . .]. The Bund, on the other hand, is a wholly conscious phenomenon derived from mutual sentiment and feeling [. . .]. (1998: 89)

In a contemporary context, what is of note about this understanding of affective-emotional sociation and action is that it reflects contemporary interest in 'affective community', and attempts to think the (apparently) unthought relation between 'community' and 'affect' (Secomb, 2003b: 9). The similarities between this perspective and the concept of the Bund are marked, but curiously not remarked upon by contemporary thinkers of affective community.

Schmalenbach 'presupposes [. . .] the individual prior to social relations, [. . .] an individual who can realise him or herself within human sociation with others' (Hetherington, 1998: 92). Accordingly, Hetherington argues that 'others—indeed, the category of Other—become of significance on emotional and moral grounds' (94). The category of the Other is understood here in a double sense: on the one hand, after Emmanuel Levinas, 'as the unknowable presence of alterity (difference)', and on the other hand, 'as strangeness and marginality' (94–95). While Hetherington's exegesis is sketchy in this regard, it is purported that 'in seeking to identify with others in a Bund-like sociation, those involved seek the Other' in both of the above senses (95). Despite the lack of detail on this point, what is of potential value in the Bund concept is that it reflects contemporary interest in the affectivity of community as a means to 'question the mythic ideal of harmonious communion' (Secomb, 2003b: 11). It is said to offer, in other words, a means 'to undo the grammar and logics of unity, to unwork and unravel totality, in the hope that strange communities, communities of strangers, may yet be to come' (11).

The Bund concept is also relevant for its potential usefulness for understanding technologically mediated forms of community. It should be noted that at no point does Hetherington address directly the translatability of Bund forms of sociation to virtual community; he merely alludes to its possible relevance in a footnote, proposing that 'it would be interesting to see if Bund-like emotional communities can exist via the Internet' (Hetherington, 1998: 100, note 4). Elsewhere, a possible connection has been recognised as having potential merit, but remains undeveloped (see Bell, 2001: 107). Nevertheless, the Bund concept can usefully illuminate understanding of (virtual) community in a number of ways. To begin with, the description of Bünde—the plural of Bund—as formed through 'mutual sentiment' accords with prevailing views of what constitutes the community of virtual community.

A further, potentially very rich, way in which the Bund concept can illuminate understanding of virtual community is via Hetherington's emphasis on the notion of 'fleetingness'. This term is again understood by approaching the Bund concept as a 'corrective' to Tönnies's twin notions of *Gemeinschaft* and *Gesellschaft* and which, as Hetherington points out, is believed by Schmalenbach to create a less rigid and more cyclical view of social change.

The key point in this cyclical model is that 'Bünde are inherently unstable and fleeting, liable to be turned into either gemeinschaftlich or gesellschaftlich forms of sociation as they break up or are routinised' (Hetherington, 1988: 90). Significantly, this characteristic of inherent instability and fleetingness is placed at the top of a prioritised list of seven points which Hetherington constructs in sketching his own understanding of the Bund concept. This first point reads as follows:

> A Bund is an elective, unstable, affectual form of sociation. This instability can be associated with its intermediate and often transitory position and character. (Hetherington, 1998: 98)

In other words, the Bund can be understood as a *liminal* form of sociation, 'a place for the expression of enthusiasms, of ferment, and of unusual doings' (Freund, 1978: 183).

In the context of CMC and virtual community, this particular conception of the Bund—as a transgressive, liminal social space—holds twofold interest. On the one hand, it has potential in negotiating and thinking through the intermingled nature of teletechnologies and community.

On the other hand, the emphasis on fleetingness in the Bund concept is useful for understanding a key and persistent criticism of virtual community. This criticism generally rehearses the following logic: bodily absence, anonymity, and the ease with which individuals can withdraw from participation in virtual community means that this form of community lacks stake and commitment. While the stock arguments for bodily absence and anonymity are already and increasingly being challenged (as was argued in the previous chapter), the issue of withdrawal is a particularly crucial one as it leads to the claim that virtual community is politically and socially impoverished (Willson, 1997; Robins, 1999; Robins and Webster, 1999). The notion of fleetingness, however, throws a somewhat different light on this latter claim. As already noted, Bünde formations are considered to be inherently unstable and 'often transitory [in] position and character' (Hetherington, 1998: 98). What this understanding suggests is that the importance of stake and commitment in community is perhaps overstated. However, within both sociology and political science, this is a potentially explosive suggestion.[8] Another way of putting this is to suggest that the issue of *electivity*—the ability to enter and withdraw (at will) from forms of sociation—has a more central place in current conceptions of *affective* community.

The twin configuration of Bünde as unstable and affective also provide insight into an often-criticised aspect of on-line social interaction: the prevalence of 'flaming' (email abuse directed at a community member or list participant). Hetherington (1998: 98) claims that 'a Bund not only establishes the conditions for identity formation, but sets this process in a form of sociation that requires intense, emotional identification with others based on a strong sense of self-governance'. It is a form of sociation that can create volatile conditions of interaction, such that 'empathic relations come to be seen as unmediated and direct, based purely in feeling', which in turn can 'provide a strong positive sense of communitas or, alternatively, bitter recrimination when things do not work out' (94).

Yet, despite the potential of the Bund concept—at least as argued by critics such as Hetherington—the efficacy of this notion for rethinking community and approaching CMC is undermined in several key ways. An initial, or general, criticism is that the Bund concept's 'intermediary status' between Tönnies's better known and established terms means it does not go far enough toward radically rethinking prevailing Anglo-American and European conceptions of community. Moreover, and despite interest in the Bund concept within cybercultural research (Bell, 2001: 107), advocates of virtual community are likely to suggest that the Bund is of limited usefulness as it is 'generally based on face-to-face interaction' (Hetherington, 1998: 98).

Extending beyond these broad limitations, however, are more specific shortcomings or difficulties, which stem from the etymology of the word Bund and its English translation. Etymologically, the term Bund, as noted above, is derived from the Indo-Germanic verb *bhend*, meaning 'to bind' or 'to tie'. It is a meaning which carries strong echoes, via the *com + munis* derivation of the word community, of community as mutual indebtedness. In the second instance—that of translation—any likely wider application of the Bund concept is severely effected by an unfortunate act of translation, where, in the most widely disseminated (albeit abridged) English-language edition of the essay, '*Bund*' has been rendered as 'communion', carrying strong echoes of the alternative rendering of community, the *com + unus* etymological formation, and thereby reinforcing an understanding of community as 'absorption into oneness'. In both cases, this legacy of meaning and translation suggests that the concept of the Bund fails to break from the 'two classic ways the West has tried to theorize community' (Van Den Abbeele, 1991: xi).

To close this examination of perspectives on community from the early twentieth century and the work of Tönnies and Schmalenbach, the following conclusions can be drawn.

With respect to the work of Tönnies, first, the distinction between the two ideas of *Gemeinschaft* and *Gesellschaft* can be read to articulate the prevailing—or at least enduring—sentiment of community as simultaneously 'lingering and decaying' (if not lost). At the same time, it emphasises community as a concept that is resilient and which carries lasting appeal. Secondly, Tönnies's alignment of *Gesellschaft* with capital and the emergence of the nation state is considered by later commentators to be crucial to understanding the desire to (re)turn to community as an idea in the mid-twentieth century.

Schmalenbach's work on the notion of the Bund is interesting for the way that it qualifies Tönnies's model by adding a third, bridging term, and therefore supplements and qualifies this influential conception of the community/society relationship.

Schmalenbach's conception of the Bund is also interesting for the promise it holds for illuminating existing understandings of computer-mediated forms of social interaction. It also parallels recent interest in the concept of 'affective community' (Secomb, 2003a; 2003b; Gandhi, 2006) in the way that it offers an affective model of sociality based on electivity, mutual sentiment, and feeling, one which is said to be open to 'the Other'. Further, it illuminates understanding of virtual community in useful ways through its emphasis on liminality and fleetingness (as a way of providing a different perspective on issues such as stake, commitment and relationship dynamics such as 'flaming'). The Bund concept reinforces (or reformulates) Secomb's earlier point that individuation is important, but so too is the need for some form of sociation that accommodates Otherness. And while commitment is undoubtedly important, the Bund concept promotes something that is far from an aseptic or unrealistically conflict-free model of community; it also offers an alternative take on the criticism that virtual community constitutes a jaundiced form of social organisation.

Nevertheless, the Bund concept's considerable shortcomings include its etymological associations and common English translation as 'communion', both of which are suggestive of community as 'fusion into oneness' and therefore promoting the exclusion of others ('the Other'). It is for this reason that the Bund concept can arguably be seen to fail to break from the two established 'Western' approaches to theorising community. This conclusion is reinforced by the fact that the 'long and convoluted history' of the term Bund saddles it with a great deal of cultural and historical baggage[9]—including appropriation by the Nazis (Hetherington, 1998: 87).[10]

Having offered these brief conclusions of the preceding material, discussion now turns to the second set of perspectives on community, in this case drawn from the mid-twentieth century.

Proliferating Definitions, and Professions as Community

In this second section, which focuses on the mid-twentieth century, two key American sociological essays from the 1950s have been selected for the way that they illuminate contemporary, particularly computer-mediated, forms and understandings of community. The first is by George Hillery Jnr, and the second by William Goode. The first essay summarises and crystallises certain ideas about community as it is commonly defined. The second examines a particular form of community that is especially useful for refining understanding about the constitutive nature of virtual community. Both essays help with addressing issues of definition in ways that are instructive for thinking about ideas of place in the context of community. Additionally, both essays have formed key sources in discussions of community and place in respect to virtual community (see, for example, Cherny, 1999). Prior to discussing them both, it is worth framing these essays by briefly situating them within a broader historical and disciplinary context.

The two essays under examination here were produced within the context of an American empirical sociological tradition that, historically, has had a strong focus on small town communities (Silver, 1990; Hummon, 1990)—and it is noteworthy in this context that Hillery's essay was published in the journal *Rural Sociology*. Furthermore, Hillery and Goode were also writing at a time not dissimilar in some respects to the period experienced by the early twentieth century writers discussed earlier. In the post-war period of 1950s America, there was relative economic stability and growth, continuing industrialisation, and urban and suburban expansion, not to mention the rise of the professional classes and 'ubiquity of the mass media' (Vidich and Bensman, 1968a). All of these factors fuelled an already existing interest in local (especially rural) community, but they also fuelled mounting anxiety about the long-term viability and health of this form of community (Warren, 1968). Explaining the impact of these changes on rural communities, particularly the industrialisation of farm machinery, one writer states:

> Tractors have largely replaced horses; and the ingenious machines associated with the tractor have so reduced the need for human labor

that migration from farms to cities, towns, and villages has become a spectacle rivalling the enclosures in Britain in the early stages of the Industrial Revolution. (Nelson, 1955: v)

These impacts and transformations lead to altered and ever-changing understandings of the notion and nature of community. From empirical studies of small town communities under these conditions a complex picture emerges where, as one commentator puts it, it is 'less possible than ever to talk realistically about a rural-urban dichotomy or even a continuum' (Bell and Newby, 1975: 116; see also Vidich and Bensman, 1968b; Dewey, 1968). Many of these impacts—such as the changing nature of community and the definitional challenges this poses, along with the rise of the professional classes—are evident in the two essays under discussion here.

As is clear from the previous chapter, the desire to arrive at definitional precision and consensus in relation to the emergent phenomena of virtual community seems to lead, almost invariably, to definitional proliferation. This is also true of community as more generally understood. Indeed, the problem of definitional proliferation is a constant in all historical attempts to think community.

This problem of the ever-expanding view of what constitutes community is documented mid-century in George Hillery's landmark (1955) study, 'Definitions of Community: Areas of Agreement'. As Hillery observes:

> Among sociologists who focus their study on community, concepts have attained such a degree of heterogeneity that it is difficult to determine whether anyone of the resulting definitions, or even any one group of definitions, affords an adequate description. (Hillery, 1955: 111)

In an attempt to make this determination, Hillery conducts a desktop study in the form of a library shelf survey of all the books offering 'explicit statements' (113) that define community. The result of this survey is the compilation of 94 different definitions. This total is revealing insofar as it bears witness to the sheer proliferation of available definitions of community, with many more texts excluded by Hillery at the time of the study because 'no clear formulation of a definition of community could be obtained from them' (113).

Equally revealing are the outcomes of comparative analysis performed on these definitions: out of the 94 examined, the only common element was that, 'all definitions deal with people' (117). 'Beyond this common basis', Hillery remarks, 'there is no agreement' (117).

Nevertheless, despite the somewhat limiting realisation that human involvement is the only common element, other areas of broad or majority agreement do emerge. Of the 94 definitions that were studied, a majority included the following elements, in order of least importance to greatest importance: geographic area, common ties, and social interaction (118).

The low importance of area in the above summary is an issue that I will return to shortly. For the moment, however, it is the other two 'elements' of community—common ties and social interaction—that are of immediate interest. This is because these two elements form the basis from and by which virtual communitarians, such as Rheingold and others, argue that computer-mediated communication constitutes a legitimate form of community. That is to say, conceiving of community as constituted by social interaction and commonality (as opposed to geographically determined criteria) is accommodating of such diverse social practices and technologies as participation in bulletin board systems (BBSs), multi-user dungeons (MUDs) and multi-user object-oriented worlds (MOOs), email lists, and so forth—all of which have been discussed as forms of virtual *community*. This 'fit' between definition and phenomena is possible because of the looseness permitted by these two terms, and perhaps speaks more of the inherent difficulties associated with a taxonomic project such as Hillery's than it does the various practices and technologies described as 'virtual community'.

It has been said that a 'primary problem of taxonomy is to determine the essential nature of the thing classified' (Jonassen, 1959: 17). In Hillery's case, the 'thing' is community, and it is the 'essence' of community that he is attempting to arrive at in his taxonomic survey. The irony, however, is that the search for the essential arrives at the general, and thus, the difficulty with the aforementioned elements of common ties and social interaction in Hillery's summary of findings is that they are *general* categories which lack analytic precision and therefore do not resolve the twin problems of definitional proliferation and the lack of definitional clarity, particularly for their potential to inform understanding of virtual community.

This is where sociologist William Goode's short but influential (1957) paper, 'Community within a Community: The Professions', can add to Hillery's findings. Goode's (1957) paper explores, 'the structural strains and supports between a contained community and the larger society of which it is a part and on which it is dependent' (194). The 'contained community' in Goode's study is the 'community of profession', such as that formed by doctors and academics.

Goode develops eight-point criteria against which professions can be measured. His argument is that, if all eight criteria are met, a profession may be considered a community. These criteria are as follows:

1. Its members are bound by a sense of identity.
2. Once in it, few leave, so that it is a terminal or continuing status for the most part.
3. Its members share values in common.
4. Its role definitions vis-à-vis both members and non-members are agreed upon and are the same for all members.
5. Within the areas of communal action there is a common language, which is understood only partially by outsiders.

6. The Community has power over its members.
7. Its limits are reasonably clear, though they are not physical and geographical, but social.
8. It produces the next generation socially through its control over the selection of professional trainees, and through its training process it sends these recruits through an adult socialization process. (Goode, 1957: 194)

There are clear parallels between Goode's criteria, Hillery's contemporaneous study of community definitions, and the idea of 'virtual community'. Each of these criteria will be discussed in turn in order to draw out these parallels—apart from item eight, from which it is difficult to draw comparisons to virtual community.

Items one and two are essentially a reiteration of Hillery's two key elements.

Items two and four, however, are decidedly more contentious in the context of virtual community research. The issue of commitment or stake, which is implicit in the second item in Goode's list, is one of the key points of contention that dominates debate on virtual community research. The common complaint is that virtual community constitutes a 'thin' form of sociation (Parrish, 2002; Kavanaugh, et al., 2005), which is to say that the ease of withdrawal from CMC—either by unsubscribing from a list or simply logging off a computer—is claimed to make virtual community politically and socially impoverished as a form of collective social interaction.

Somewhat less contentious is the issue of a common language only partially understood by outsiders, as contained in item five in the above list. Indeed, analysing the discursive patterns and social codes that make up a 'common language' has been a central strategy for Internet researchers. This is most evident in what David Bell describes as the many 'talk-and-text' analyses of virtual community, especially those linguistically oriented studies which examine the various communicative strategies and practices which make up the language of CMC, such as the use of emoticons (diacritical marks employed to simulate facial expressions and emotions), as well as other protocols of 'netiquette' (including minor transgressions of them, such as lurking, cross-posting, 'shouting', 'flaming', etc.) (Bell, 2001: 102–103). While this approach is not taken up or discussed in detail here, there are two useful points to remember. First, this approach to CMC study is part of a continuum of community research examining the various forms, codes, and practices which shape and give character to 'contained communities'. Secondly, and underlying the preceding point, the fact that many forms of CMC *do* possess a common language only partially understood by outsiders goes some way towards confirming these on-line practices as a form of 'community'.

This confirmation is furthered by considering item six in Goode's list: the notion that community has power over its members. Goode discusses this issue primarily in terms of protectionism within professions, in the sense that 'insiders will generally fare better than outsiders'. But it is also meant in terms of social control, where, 'in exchange for protection against the larger society [including punishment for wrong doing], the professional accepts the social control of the professional community [to which s/he belongs]' (Goode, 1957: 198). This is perhaps clearest when one thinks of historical guilds and unions, as well as the sorts of professional associations of which Goode writes (such as those for doctors, lawyers, architects, and even real estate agents).

But the issue of power is also connected to identity, in the sense that membership of a professional community not only bestows a certain level of professional prestige on an individual member, it can also bring a sense of identity as well. In fact, these two understandings of power come together in membership, which, as has been noted in other studies of community, not only 'involves the conscious recognition of articulated interests in common,' it also 'implies both *participation* and *authority*' (Plant, 1974: 51). In this context, authority refers to community as associated with, and functioning within, some form of rule-governed interaction:

> However much the language of community work may talk in terms of unstructured situations, groups and communities[,] this cannot, if these situations and groups and communities are to be determinate things at all, mean that such groups are without rules and thus without authority. (Plant, 1974: 57–58)

This is not a rejection of 'rules and authority in itself', but rather a 'not very precise protest against the complexity of authority' (57–58).

It is true that this understanding of community membership as implying authority has undergone strong challenge in recent times, as has the notion of community as 'work'—both of which will be explored later in this chapter. Nevertheless, the above passage points to an important fact: the forms of on-line sociation that are commonly grouped under the rubric of 'virtual community' are, for the most part, not generally anarchistic. That is to say, virtual communities are rule-governed and exert power over members in similar ways to contained professional communities. This is experienced both through general adherence to accepted, if unwritten, informal protocols of 'netiquette' (notwithstanding already mentioned practices such as 'flaming'), and, more formally (if sometimes on an ad hoc basis), through list 'moderation', and through the formation of 'closed' lists. (Attempts to operate outside these protocols are well-documented in the literature, but are generally short-lived.) Such 'authority' and participation in rule-governed social interaction is important in that it further suggests that virtual community reflects more conventional models of community.

To conclude this discussion of mid-twentieth century perspectives on the notion of community, the following observations can be made in light of the work of Hillery and Goode.

The first is that Hillery's survey reveals that the most important and recurrent elements in definitions of community (at least up until the mid-1950s, when his study was conducted) are people, area, common ties, and social interaction. The fact that the last two criteria—common ties and social interaction—are considered to be of greatest importance is significant in light of the findings of the previous chapter on virtual community. What emerged there is that these same two criteria of common ties and social interaction recur in and are considered fundamental to definitions of virtual community. This is valuable insofar as it emphasises a degree of continuity between existing understandings of the notion of community, and 'emergent', computer-mediated forms of social interaction and community. However, as noted in the conclusions to the previous chapter, this convergence or 'fit' between established and computer-mediated forms of community also poses manifold issues for the future of virtual community as a concept and as a distinct form of community and whether it can continue to be conceived as such.

Goode's examination of the professions adds to these findings by repeating Hillery's emphasis on the importance of common ties and shared interest. It also does so through its emphasis on a formulation of community as plural and layered, with 'contained communities' existing within and dependent on a larger society (Goode, 1957: 200; see also Warren, 1968: 193–200). While this delineation perhaps seems commonsensical, it is in fact a particularly important insight, and touches on an underlying concern of this book: how to move beyond the common (mis)conception, especially in 'early' and boosterist accounts of virtual community, of these on-line social spaces as detached, or separate, from other community and wider society. While such separatist visions are undergoing increasing challenge, it is a polarity that lingers. And, by way of response, Goode's conception of community as plural and imbricated offers a model that is both simple and sensible. Moreover, it forms a useful basis from which to come to terms with the intermingled nature of the 'real' and the 'virtual'—an issue that will be addressed further in the final chapter of this book.

The second observation returns to Hillery's study. Of equal interest to those elements considered to be most crucial to definitions of community is the idea that 'area' (or local 'place') is considered the least crucial element in definitions of community. This finding is certainly supported by Goode's study. Area barely figures at all in his focus on the professions as forms of geographically dispersed social aggregations. This also underscores Doreen Massey's (1991: 24) point that '"place" and "community" have only rarely been coterminous'.

The reduced importance that is placed on area in definitions of community is particularly significant in light of virtual community research which argues that a key aspect of virtual community is that it operates independent of area (and geography). Hillery's finding qualifies this claim, and demonstrates that this desire to separate community from place has been sought (and has been documented) for many decades previous to interest in networked teletechnologies like the Internet.

However, in noting the comparative marginality of 'area' as a core constitutive element of most definitions of community, Hillery stresses that, while 'area' (or local 'place') may not be the most important definitional element, this does not preclude it from definitions of community. He writes:

> In claiming that area [. . .] is not necessary for community (which some authors do), no claim is made that a group which *is* in a specific area *cannot* be a community. (Hillery, 1955: 118)

The non-constitutional nature of 'area' is one reason that some authors insist on the semantic delineation between 'the community' and 'community'. The former is taken to refer to 'settlements of the kind encompassed in definitions [. . .] in which *locale* is a basic component' (Bernard, 1973: 3). For example, in the introduction to their book *Community Studies*, Colin Bell and Howard Newby write that:

> A community study must be concerned with the study of the *interrelationships* of social institutions within a locality. This does not have to mean that *all* social institutions have to be studied but, unless these *interrelations* are considered, they will not be considered community studies for the purposes of this book. (Bell and Newby, 1975: 19; see also König, 1968)[11]

More common is the second position, which, in accordance with Hillery's findings, 'emphasizes [. . .] common-ties and social interaction' (Bernard, 1973: 4).[12]

Moreover, and crucially, to say that 'area' (or local 'place') is the least crucial element in most definitions of community (including 'virtual community') is not to say that area and place are irrelevant or unimportant to understandings of community (see Hummon, 1992),[13] including those forms that are enabled by networked teletechnologies (Wellman, 2001: 41, note 1). Rather, one of the key concerns of this book is to see how 'place' (which, as understood here, includes but is not restricted to the idea of 'area') relates to and informs understanding of social interaction and 'community'.

In summary, then, these perspectives on community from mid-twentieth century sociological research serve to validate 'virtual community' as a legitimate form of community. At the same time, these perspectives demonstrate that conceptions of and arguments about CMC and 'virtual community' can and perhaps need to be situated historically within a continuum of community research.

The following, and final, section of this examination of historical perspectives on community turns to the late twentieth century, and from sociological literature to the context of continental philosophy.

Beyond Community

In examining perspectives on the notion of community from the late twentieth century, there are a number of different approaches that might be selected from for closer examination. For example, there is the work of the 'communitarian' philosophers (Alisdair MacIntyre, Michael Walzer, Michael Sandel, and Charles Taylor), the work of the so-called 'ideological communitarians' (especially those involved with the 'Responsive Communitarian Platform' in the US, including Amitai Etzioni, Mary Ann Glendon, and William Galston), and even those writers working at the forefront of a revival in questions of community within the social sciences and contemporary sociology (such as Zygmunt Bauman and Gerard Delanty).

However, the approach taken here is to concentrate on the work of two French poststructuralist philosophers, Jean-Luc Nancy and Jacques Derrida, since the work of these two writers has increasingly become central to debates about the idea of community (Ahmed and Fortier, 2003; Glowacka, 2006). There are a number of reasons for this selection. First, the approach to thinking about community that is taken by Nancy and Derrida differs markedly from the approaches examined thus far, and thus provides a useful point of contrast. Secondly, both thinkers question the foundations on which community has hitherto been conceived, by rethinking the very idea of community. Thirdly, both attempt to move beyond these limitations by developing models of community that offer a striking departure from more traditional conceptions of community, and which specifically address issues of exclusion and of how to accommodate difference and 'the Other'. For example, it has been said of Nancy—and the same is true of Derrida—that his work, 'is deliberately attempting to stretch and rework concepts such as community to find new ways to discuss and understand' such phenomena (Willson, 2006: 147–148). The result of 'this pushing of concepts to their limits' is that it 'makes any textual analysis difficult, since the work is full of contradictions, clarifications and overlap' (148). These difficulties notwithstanding, isolating the key points of both thinkers critique of community is valuable insofar as it provides an overview of an increasingly influential alternative to more traditional 'Western' historical conceptions of community.

Nancy and Derrida (along with Maurice Blanchot and Emmanuel Levinas) share a similar philosophical heritage (Morin, 2006; Korhonen, 2006). As Michele Willson (2006: 147) notes, 'these theorists became prominent within academia after the revelations of the Holocaust, and the disaster/demise of the practice of communism in Eastern Europe'. In the context of these events, both Nancy and Derrida are interested in thinking through the perils that attend 'an understanding of community based around a singular notion of identity, of totalitarianism, and of undertaking projects of community that became apparent as a result of these events' (Willson, 2006: 147).[14] In addition, both have been influenced in key ways—both directly and indirectly—by the work of Martin Heidegger (Willson, 2006: 147), whose ideas and writings on the nature of 'being' (ontology) permeate much of twentieth century philosophy generally and continental philosophy in particular.

With this background in mind, the following section examines the approach that is taken by Nancy and Derrida to rethinking community. Discussion begins with Jean-Luc Nancy, who has developed one of the most striking departures from more traditional, dialectical ways of conceiving of the problematics of community in *The Inoperative Community* (1991b) and other writings.

Nancy is particularly suspicious of various theoretical attempts throughout Anglo-American and European history centred upon recapturing community (Willson, 2006: 148; Nancy, 1991b: 1–10). His suspicion is directed at the Christian understanding of community as communion—or what Nancy (1991b: 10) comes to term community's 'onto-theological' heritage. This conflation of community as communion, it is claimed, is 'more recently translatable within modernity as the desire for immanence and a defying of the finitude that is the reality of each human being' (Willson, 2006: 148). For Nancy, community is not communion. This is because community as communion, 'constrains a group of people into a monolithic form or identity, suppressing difference and promoting exclusionary practices' (Willson, 2006: 149).

In contradistinction to this understanding of community as communion, Nancy suggests that the 'ultimate limit of community, or the limit that is formed by community [. . .] traces an entirely different line' (1991b: 8), one that is concerned with the failure of community as an idea. Nancy argues that this failure is due to an embrace of the notion of 'immanence'—that is, of totality, self-consciousness, self-presence—which itself is anchored in a ('Western') metaphysical tradition that has 'largely been unable to think without recourse to the subject' (Van Den Abbeele, 1991: xiv; Nancy, 1991b: 14). Nancy's reworking of community, Georges Van Den Abbeele (1991: xiv) explains, 'is neither a community of subjects, nor a promise of immanence, nor a communion of individuals in some higher or greater totality (a State, a nation, a People, etc.)'. More particularly, community 'cannot arise from the domain of work', it is not an *oeuvre*, and 'one does not produce it' (Nancy, 1991b: 31). Rather, for Nancy (following Bataille) community is what is '"unworked" (*désoeuvré*)'—'one experiences or one is constituted by it as the experience of finitude' (1991b: 31; Van Den Abbeele, 1991: xiv). As Ian James explains:

> What is being suggested here is that community is not and never has been possible on the basis of an intimate and totalized sharing of an essence or identity, which might then be lost, ruptured, or dispersed and that we might long to regain. Rather community is possible, in the first instance and on a primordial level, only as a kind of rupturing or dispersion, which is itself constitutive of the sharing

or communication proper to the being-in-common of the communal. (James, 2006: 175–176)

Despite the opacity of Nancy's ideas, and the overlap between them, it is possible to isolate from the above discussion several key concepts or themes, disentangling and elaborating on each.

The first concerns Nancy's notion of 'being-in-common'. Nancy 'argues for an understanding of community as the *incomplete* sharing of the relation between beings, where there exists sharing and separation or resistance of beings simultaneously' (Willson, 2006: 149). As he puts it, 'there is no communion, there is no common being, but there is being *in* common' (Nancy, 1991c: 4). This is to say:

> being is not common since it differs with each experience of existence, but being is *in* common: It is in the *in* where community 'resides'. Community exists in/as a relation between being, existence [. . .], and finitude. (Willson, 2006: 149)

This relational conception of community—and the emphasis on finitude—can be further explained by considering a second key Nancean notion: that of community as '"un-worked" (*dés-oeuvré*)' (1991b: 31). The concept of a 'workless' community is indebted to the writing of Georges Bataille, and has been developed and reformulated by a generation of philosophers both contemporary with and following after Bataille (including Nancy, as well as Blanchot and Derrida). The notion of a 'workless' community pivots around 'the Heideggerian thematics of finitude', defined as 'the acutely temporal sense in which a being is oriented toward death' (Lee, 1998: 68, note 5).[15]

This is so, it is suggested, because 'workless' communities are organised around each member's finitude or loss:

> Such communities are described as 'workless' because its members are not brought together through a shared work, project, or set of interests, or lived experiences. Rather, it is the mutual recognition of the finitude or radical otherness of its members, as specifically witnessed in the other's passing, that is the foundation of the workless community. (Lee, 1998: 68, note 5)

And because 'community is not a *thing* that can be consciously created or artificially imposed' (Willson, 2006: 149), the idea that community is formed through bonding or commonality is problematic for Nancy.[16] The reason: 'if community is encapsulated, then it is lost in that process, for it cannot be reproduced or fixed [as happened during Nazi rule] as it differs with each occurrence or presentation' (Willson, 2006: 150; see also Nancy, 1992).

Such a formulation would appear to present certain problems. If community is 'singular' and cannot be reproduced or fixed, how, then, might Nancy's formulation address what was described earlier as the 'problem of community'? That is, how can participants be-together-in-community and simultaneously remain open to the other? Nancy's response is to try to 'overcome the integrative/differentiating dilemma faced by community theory through asserting the impossibility of either individuals without relations, or of collectives without singularities, through the notion of singular beings' (Willson, 2006: 150). Thus, for Nancy, 'community exists in/on the limit where singular beings meet' (Willson, 2006: 151). He writes, 'Being cannot *be* anything but being-with-another, circulating in the *with* and as the *with* of this singularly plural coexistence' (Nancy, 2000: 172).

These intertwined notions of limit and singularity together form the third key concept in Nancy's writing. As Willson explains:

> Singular beings are not fixed totalities: They exist in and through their relations with other singular beings. [. . .] A singular being ends at the point where it meets other singular being/s. This is the *limit* where the possibility of one or another singular being exist simultaneously. The limit refers to a mutual exposure whereby both presence and absence exist simultaneously. (Willson, 2006: 151)

If there is a unifying theme to Nancy's work, then, it is the project of deconstructing the subject, through a shift away from community and toward a recognition of political passion (Nancy, 2003; 1991a). It is a movement involving risk and potential:

> Passion embraces both the experience of suffering and of desire. To recognize a passion that destabilizes the unity of self means experiencing oneself as different and uncommon. The disconcerting task that ensues is to endure the affect that feels like a strange self, and the risk is to expose oneself to others in such an unaffected non-selfsame state. (Rauch, 1996: 253)[17]

It is the risk inherent in such exposure that further informs the Nancean understanding of thinking at the limit of community (Kamuf, 1991). Indeed, it is in thinking the risk of the limit—or thinking the risk that takes place through exposure of/at the limit of community—that Nancy is committed to developing an ethics of and for the other.

> His understanding of incomplete sharing stresses both the relational and also the resistant nature of being in the world. Such an assertion about the necessity of the relational acknowledges the uniqueness of other beings and imposes an implicit ethical responsibility to allow and respect differences. (Willson, 2006: 151)

Herein lays the greatest strength of Nancy's theoretical speculations on community. As Willson (2006: 173) argues, 'perhaps the most important aspect of Nancy's work is his struggle with the tensions of singularity and commonality, of integration and difference.' What it reveals, however, is that no absolutely satisfactory resolution can be achieved: 'The struggle is interminable and unresolvable' (Willson, 2006: 173), but is not one that should be abandoned because of this.

Having sketched the broad contours of Nancy's theoretical work on the concept of community, it is valuable at this point to consider some criticisms of these theories.

In her detailed critical engagement with Nancy's work, the Australian Internet scholar Michele Willson raises a number of issues in relation to Nancy's attempts to reconceptualise community that warrant attention here. To begin with, Willson (2006) notes the difficulties facing any attempt to apply or transpose Nancy's theories onto the sorts of communicative practices that characterise on-line or virtual community, both because 'both forms or communicative intents (work and unwork [or 'play']) take place' (170) within CMC, and as a result of his lack of a clear statement on teletechnology use. On this latter issue, Willson suggests that from an engagement with Nancy's work, teletechnologies could be seen to have both potentially negative connotations for the way that they may be employed instrumentally and in terms of 'work' (Willson, 2006: 168), while 'simultaneously' opening up 'the possibilities for infinite communication enabled by such technological forms' (169). Hence, the crucial point, which applies to the aforementioned 'communicative intents' of CMC as well, is that in Nancean terms it is not technology per se that is likely to be an issue, but rather specific engagements with it that lend support to and reinforce restrictive forms of unity.

A second issue relates to the political implications of Nancy's emphasis on 'workless' community. As already noted, Nancy insists that community cannot be 'made', that it 'cannot arise from the domain of *work*':

> One does not produce it [. . .]. Community understood as a work or through its works would presuppose that the common being, as such, be objectifiable and producible (in sites, persons, buildings, discourses, institutions, symbols: in short, in subjects). (Nancy, 1991b: 31)

The difficulty with this withdrawal from work, Willson suggests, is that it can lead to the charge that this is also a withdrawal from politics and the realisation of practical change, and a 'retreat' into the realm of abstraction (Willson, 2006: 154–155). This charge is refuted elsewhere. For instance, Fynsk (1991) argues that Nancy's writing contributes to a political reinvigoration of philosophy—a counter-argument which is admittedly acknowledged by Willson in a footnote (2006: 174, note 12). A similar argument is developed at length by James (2006: 152–193), who sees Nancy's ideas as 'resonat[ing] consistently with profoundly political implications' (153) and his writing as offering 'an important rethinking of the relation between the philosophical, the political, and politics' (154). Indeed, in a revealing later reflection on *The Inoperable Community*, a short piece written in October 2001 in the immediate aftermath of the events of September 11, Nancy (2003) restates the political importance of philosophical reengagement with, and caution with respect to, the concept of 'community' and all that this word implies.

A third concern with Nancy's work also relates to his insistence on 'workless' and 'unstable' forms of 'community'. Paraphrasing Nancy, Boyarin and Boyarin (1993: 698) write that, 'the only community that does not betray the hope invested in [the word *being*], is the one that resists any kind of stable existence'. From this summary, the two authors then proceed to the following erroneous conclusion: 'Any already existing "community" is out of consideration by its very existence, relegated through philosophical necessity to a world we have lost or that never existed' (Boyarin and Boyarin, 1993: 698). It is erroneous as Nancy is quite clear on this point:

> there has been, already, always already, a 'work' of community, an operation of sharing out that will always have gone before any singular or generic existence [and . . .] without which it would be unthinkable to have, in an absolute general manner, any *presence* or any *world*. (Nancy, 2003: 32)

Nancy (2003: 32) does not deny an already established 'co-existence' or 'co-belonging'. What he does argue for is the need to rethink this 'belonging' in specific terms as 'no more than a belonging to the fact of being-in-common' (32).

Finally, and in addition to the above point, Willson picks up on the phrase 'being-in-common' and suggests that, while Nancy 'continuously asserts the singularity (uniqueness) of each/all experience, [. . .] it could be argued that, if being is *in* common, then surely there must be some thing, some essence, quality, procedure, or element that is shared—even if only incompletely—that enables Nancy to say that "being is *in* common"' (Willson, 2006: 162–163). Again, Nancy would not dispute this. In his aforementioned 2003 article, Nancy relates how Blanchot wrote *The Unavowable Community* (1988) as a response to his own earlier text, *The Inoperative Community*; it is a response which, Nancy reflects, prompted him to continue to examine 'the case of the "common"—but also the enigma or the difficulty of it, its non-given, non-available character, making it, in this sense, the least "common" quality in the world' (Nancy, 2003: 30). As Nancy notes elsewhere:

> Community [. . .] is neither an abstract or immaterial relationship, nor a common substance. It is not a common *being*; it is to be *in* common, or to be *with* each other, or to be together. (1993: 154)

Nancy's later preference for the word 'with'—'I have preferred to substitute, little by little, the graceless expressions, "being-together", "being-in-common", and finally "being-with"' (Nancy, 2003: 31; see also Devisch, 2006)—is illuminating here in that it further clarifies his philosophical position. His preference for the 'with' in 'being-with' is that 'it brings with it [. . .] a clearer indicator of the removal at the heart of proximity and intimacy', a removal which emphasises 'neither communion nor atomisation' (32).

Having detailed (and responded to) a number of criticisms of his work, it is valuable to conclude this discussion of Nancy by restating the most significant feature of his work. As noted above, one of the greatest strengths of Nancy's theoretical speculations on community is 'his struggle with the tensions of singularity and commonality, of integration and difference' (Willson, 2006: 173); that is, the desire to reconceptualise 'community' as 'being-with'—a concept which emphasises the 'with' of a 'singularly plural coexistence' (Nancy, 2000: 172).

Discussion now turns to the work of Jacques Derrida. According to Morin (2006), in his engagement with community, Derrida pursues a 'similar project' to Nancy, albeit one that is pursued via 'different strategies' (Morin, 2006). The differences will be sketched later, but in terms of similarities, there are many. Like Nancy, Derrida questions the word community because he sees that its etymology links it to violence and exclusion (see Derrida, et al., 1997; Caputo, 1997). Similarly, he prefers the word 'sharing', albeit developed in quite particular ways that will be detailed below (Derrida and Stiegler, 2002: 66). And, like Nancy, Derrida's thinking on sharing and the social occurs 'on the limit' (Kamuf, 1991), testing and stretching the many theoretical boundaries and possibilities of a reworked understanding of 'community' through his extended (and intertwined) meditations on the concepts of 'forgiveness' (Derrida, 2001b), 'friendship' (Derrida, 1997b), 'gift-giving' (Derrida, et al., 1994; Derrida, 1992, 1995b), and 'hospitality' (Derrida and Defourmantelle, 2000). However, unlike Nancy, Derrida does comment directly on the implications of teletechnologies for theories of community.

Despite Derrida's own self-distancing *vis-à-vis* the word community, attentive readers of Derrida's work have found sufficient 'give' or room for manoeuvre to critique community 'from within community', without abandoning the term altogether (Rapaport, 2003: 140–141, note 1). In fact, the claim that Derrida 'abandons' community is arguably overstated (see Rapaport, 2003: 141, note 1).

Astute critics of Derrida's texts have thus been able to isolate in and extrapolate from the texts of 'early Derrida' the beginnings (or threads) of key ideas, such as 'gift-giving'. These ideas have become more fully formed in the so-called 'ethical turn' of the texts of 'later Derrida' (Rapaport, 2003: vii), being used to dramatically rethink community.[18]

One such critic is the political scientist William Corlett (1993) who draws from Derrida's texts to argue that community is best advanced within a politics of difference.[19] According to Corlett, a mutual appreciation of differences not only problematises conventional binary oppositions (individual/communion, self/other), it displaces the centrality of the subject and 'open community' in favour of a model of 'community without unity', one that is based neither on individualism or collectivity (in the sense of fusion into oneness) but on free gift-giving and sharing (Corlett, 1993: 209–217; Delanty, 2003: 138). It is through the act of giving, moreover, that the identity of the giver and receiver is determined:

> Supposing that a gift has been given; that supposes that before it took place, the giver is not determined, and the receiver is not determined. But the gift determines; it is the determination, it produces the identity of the giver and the receiver. [. . .] That is why the gift is always a strike of force, an irruption. (Derrida, 1987: 199)[20]

The key to this understanding of gift-giving is that 'gifts cannot be calculated and are by definition accidental' (Corlett, 1993: 215).

> The gift must be given by chance. If the gift is calculated, if you know what you are going to give to whom, if you know what you want to give, for what reason, to whom, in view of what, etc., there is no longer any gift. (Derrida, 1987: 198)

This open-ended gift-giving connects Derrida's (and by extension, Corlett's) notion of the gift with Nancy's conception of the workless community; and both concepts are derived and developed from the work of Georges Bataille. Writing on Bataille's involvement in the Acéphale group (see endnote 15), Maurice Blanchot (1988: 15) quotes Bataille, who writes that, 'to sacrifice is not to kill but to abandon and to give'. Thus, as Blanchot understands it, 'to link oneself with Acéphale is to abandon and to give oneself: *to give oneself wholly to limitless abandonment*' (1988: 15—original emphasis). In a footnote, Blanchot further states that this is a 'giving without any return in mind' (1988: 57, note 6). 'In other words,' as Sheri Hoem (1996: 50) remarks, 'it is a community of those sacrificing themselves for no purpose, no new form of society, no "work", no object, that would be an end result.' This context has proven influential and is critical not only to Derrida's idea of gift-giving, but also to Nancy's preference for unstable forms of 'community'.

Derrida's notion of 'hospitality' is similar in many key respects to his conception of gift-giving without return. Both carry a certain risk and both are founded on indeterminacy (in the sense of nothing being predetermined).

> Absolute hospitality requires that I open up my home and that I give not only to the foreigner [. . .], but to the absolute, unknown, anonymous other, and that I give place to them, that I let them come, that I

let them arrive, without asking of them either reciprocity. (Derrida and Dufourmantelle, 2000: 25)

In contrast to this, Derrida writes that:

A cultural or linguistic community, a family or a nation, cannot fail at the very least to suspend if not to betray this principle of absolute hospitality: so as to protect a 'home', presumably, by guaranteeing 'one's own' against the unrestricted arrival of the other. (Derrida, 2005: 66)

Postcolonial theorist Leela Gandhi's thinking has been informed by Derrida's insights as well as the Epicurean notion of *philoxenia*, or 'love for guests, strangers, foreigners'. She suggests that this ethics-as-hospitality involves 'opening the door' to the foreigner/absolute other. Gandhi argues this is 'existentially profound', 'involving as it does the potentially "agonising" risk of self-exile which haunts any ethical capacity to become (to suffer oneself to become) foreign to "one's own", and, above all, to oneself' (Gandhi, 2003: 18–19). For, as Derrida observes of the stranger's entry:

Crossing the threshold is entering and not only approaching or coming. Strange logic, but so enlightening for us, that of an impatient master awaiting his guest as liberator, his emancipator. It is *as if* the stranger or foreigner held the keys. (Derrida and Dufourmantelle, 2000: 123)

In addition to this opening up of self and/to other, Derrida's conception of hospitality—particularly as an alternative to community—is seen by him as connected strongly with global telecommunications technologies. For example, in *Echographies of Television*, Derrida suggests that the defamiliarising effects of globalisation and networked telecommunications technologies leads to a growing and renewed desire for the 'home', in both its domestic sense and in a more threatening national sense (discussed in detail in Chapter 6) (Derrida and Stiegler, 2002: 79–80). Again, 'hospitality' carries this threat within itself. But at the same time that it carries this threat, hospitality also 'opens a door'; it contains the promise of something else, a potential countering perspective or possibility:

Without [. . .] having to give up singularity, idiom, and even a certain-at-home, this at-home which [. . . projects an image . . .] of closedness, of selfish and impoverishing and even lethal isolation, [. . .] is also the condition of openness, of hospitality, and of the door. (Derrida and Stiegler, 2002: 81)[21]

However, beyond this promise of hospitality—or the *promise* of the promise of hospitality—this projection of openness remains just that: a projection. That is to say, it exposes Derrida (like Nancy) to the charge that his formulation is explicitly 'philosophical' offering 'no concrete examples' of the terms it seeks to theorise (Delanty, 2003: 138). In other words, it presents a formulation of 'community in the abstract' (Willson, 1997). Even so, the purely philosophical nature of this model is also explicable in other, more productive ways. For example, abstraction is something of an inevitability when one is engaged in thinking 'on the limit', as both Derrida and Nancy are. Secondly, the fact that these theories of free gift-giving and hospitality remain open-ended, merely sketched, is not so much a criticism as a description of its 'project' (with this term intended in a double sense, of ideas in process, and as a 'projection' into the future, as well as in the sense of Derrida's thinking being at the time a 'work in progress'). As Derrida puts it:

> It's better to let the future open—this is the axiom of deconstruction, the thing from which it always starts out and which binds it, like the future itself, to alterity, to the priceless *dignity* of alterity, that is to say, to justice. It is also democracy as democracy to come. (Derrida and Stiegler, 2002: 21)

In other words, a 'community' that practices a politics of difference, that is constituted through free gift-giving or hospitality, that remains open to the other, is (for the present at least) a 'coming community', to use Agamben's (1993) formulation: a community deferred, a community yet-to-come, a community of and for the future, a community *in potentia* (see also Wilken, 2010: 459–463). This is precisely why Nancy's concept of 'inoperable community' (*'la "communauté désoeuvrée"'*) has also been translated elsewhere as 'unoccupied community' (Nancy, 2003: 30).

It is possible (and in some ways productive) to relate an understanding of 'community' (or its alternative) as free gift-giving and hospitality to a study of virtual community, especially as an 'economy of the gift' constitutes an important theme in the history of computer-mediated communications. For example, Peter Kollock (1999) discusses gift-giving in the context of what he terms 'economies of online cooperation' and notes that, while this practice as classically defined does occur on the Internet (such as the exchange of information between colleagues), 'much of the help and sharing that occurs is actually different from traditional gift exchange':

> When people pass on free advice or offer useful information, the recipient is often unknown to them and the giver may never encounter the recipient again. Thus, the usual obligation of a loose reciprocity between two specific individuals is difficult or impossible. Indeed, gifts of information and advice are often offered not to particular individuals, but to a group as a whole. (Kollock, 1999: 222)

An important point is missed in the above passage, however. While on-line gift-giving may occur on a scale that makes reciprocity impossible, giving (such as the sharing of information) is not done entirely without return. As the phrase 'economies of online cooperation' makes clear, this occurs within an economic system, and the 'currency' of this economy involves an exchange for which the giver, in this instance at least, receives increased symbolic or cultural capital. Thus, while on-line gift-giving has some resonance with Derrida's ideas, it nevertheless stops well short of the kind of free gift-giving without exchange or return that characterises his understanding of the gift.

The twin notions of gift-giving and hospitality (not to mention the politics of friendship) as a move beyond 'community', have also been developed productively as a way of making sense of what has been described as the postmodern (re)turn to ideas of utopianism, with the latter term meant in a very precise sense, to refer to 'a politics of alternatives poised at the limits of thought and being, epistemology and ontology'—a 'politics' that requires a 'careful renegotiation with ideas of community, communication, sociability, *conatus*' (Gandhi, 2003: 12).

This is essentially the basis from which Derrida himself frames his meditations on gift-giving, hospitality, and the 'geopolitical effect and density of telecommunications networks', especially as they impinge on and shape issues such as European nationalism, immigration, and border control (see Derrida and Stiegler, 2002). And it is in the course of these meditations that Derrida—with the customary qualifications and reservations, which are 'full of aporetic formulations, negations and denegations' (Morin, 2006)—comes closest to a formulation or description of something approximating virtual community. Reflecting on the political implications of, and sense of 'telepresence' generated by, many people in geographically dispersed locations watching the same CNN broadcast, Derrida discusses how 'network', or better still, 'sharing', are preferable terms than 'technological community':

> Like Jean-Luc Nancy, I prefer the word sharing: it both says what it is possible up to a point to have in common, and it takes dissociations, singularities, diffractions [. . .] into account. It doesn't signify a community, if by community one understands a unity of languages, of cultural, ethnic, or religious horizons. There is indeed a form of coinscription in space, or with a view to space, which doesn't correspond to the same models as before, but I would hesitate to call this a community [. . .] because it is made from different places, with different strategies, of different languages, and respect for these singularities seems just as important to me as respect for community. (Derrida and Stiegler, 2002: 66)

Like Nancy, Derrida has been praised for disconnecting communal identities from demarcated geographical spaces.[22] However, I would argue that this important passage reveals a number of insights that suggests a far more ambivalent position on this score, which might also be relevant to the role of place in virtual community discourse and analysis. To begin with, while Derrida asserts that globalisation processes have led to 'radical delocalization', there is a countervailing admission here that, indeed, some 'form of coinscription in space, or with a view to space' occurs within networked telecommunications (which I here take to include CMC). And, it would appear from the context of the passage that the 'spatial' to which Derrida refers implicates both 'actual' space and 'virtual' space, which would make sense given Derrida's stated preference for an understanding of intermingled spaces, as captured in the portmanteau word 'actuvirtuality' (Derrida and Stiegler, 2002). Given this, it is important to remember that, for Derrida, the interconnection (or at least *sense* of interconnection) between disparate places—both 'actual' and 'virtual'—that is enabled by CMC and other telecommunications technologies does not create community (at best it is 'sharing'). This is because different places involve different strategies. And, furthermore, to restate a key point of Derrida's, 'respect for these singularities seems just as important [. . .] as respect for community'.

Poststructuralist continental philosophy, as reviewed here in the work of Nancy and Derrida, thus offers one of the more radical challenges to existing conceptions of community. This challenge comes from the ability of Nancy and Derrida to make clear the existing limitations about how we, particularly in 'the West', think about the notion of community. It also comes from their attempts to move beyond these limitations and rethink community in ways which are accommodating and welcoming of difference and otherness.

Nancy's approach is to rethink community as 'workless community'. It is an understanding which emphasises the importance of 'singular beings', and 'being-*in*-common', or 'being-with'.

As noted earlier, Derrida pursues a similar project to Nancy in his own critique of community. However, Morin (2006) cautions against seeing these two projects as 'different ways of achieving the same end', remarking that 'the choice of a strategy is never neutral or meaningless'. Thus, while Nancy retains some form of engagement with the concept of community, Derrida avoids the term community altogether. Yet, as explored above, despite these aversions it is still possible to discern some sort of conception of 'community' or approach to 'community' (that is nevertheless not community as traditionally understood) sedimented in his work, especially in his preference for and pursuit of multiple trajectories that explore alternative concepts, including 'hospitality', 'gift-giving', 'friendship', 'sharing', and so on.

What these different strategies mean is that each thinker brings their own particular inflection or emphasis to a reworked understanding of 'community': Nancy insists on a fundamental type of being-together, while Derrida emphasises 'respect for the singularity and alterity of the other'

(Morin, 2006).[23] These differences notwithstanding, the work of Nancy and Derrida is thus equally significant for the way that they have addressed and developed alternatives to the persistent problem of community as communion. In the work of both, one might say, 'the fate of the Other and otherness becomes an index of the possibility of meaningful togetherness' (Holmes, 1997: 16).

Derrida's work is also valuable in the context of the present study in that, like Nancy, he remains cautious about the potentially deleterious impacts of teletechnologies, he nevertheless engages at length with the spatial, social, and political implications and possibilities of these technologies.

A further and final point to note is the extent to which, and determination with which, Nancy and Derrida both strive to separate community from place, albeit implicitly rather than explicitly. Such a desire is understandable when one recollects the context in which they are writing and to which they were responding (that is, the revelations of the Holocaust and the disaster/demise of European communism, etc.). What this would seem to demonstrate is that, while 'area' (or local 'place') has long been considered a marginal element in definitions of community, Nancy and Derrida react to a lingering sense of connection between 'place' and a particular conception of restrictive unity, which they see as troubling.

COMMUNITY AS FRAUGHT; COMMUNITY RETHOUGHT

So, from this process of wide-ranging review, what is the composite picture of community and how is it useful for discussions of teletechnologies and computer-mediated social interaction? What emerges from this discussion is a picture of community as paradox and problem. The notion of community is a paradox because it is characterised by an ongoing cycle that turns on a perceived loss of community and a yearning for its renewal. The paradox presents a problem, however, when one approaches the issue (or category) of 'the other'. That is to say, enduring Western accounts of community fail at 'remaining open to the other' (Lucy, 2001: 146) while, by the same token, recognising that 'community is nonetheless indispensable' (Secomb, 2003b: 9).

In light of these preliminary observations, and in order to develop a fuller understanding of the notion of community, a series of three sets of perspectives on community from three different points throughout the twentieth century were examined. Each of these three sets of perspectives has added to general understanding of the notion of community and to 'virtual community' in particular.

For example, the work of Herman Schmalenbach—and particularly as it has been interpreted by more recent commentators—can be seen to hold considerable promise for adding to existing understanding of CMC and virtual community. This promise is located in the flexibility of

Schmalenbach's understanding of the notion of the Bund; the way that this notion can be extrapolated to other contexts and social structures such as computer-mediated social interaction; the way that Schmalenbach sees the structure of the Bund as constituted by both more permanent and temporary or fleeting forms of social interaction; and insofar as it acknowledges that social interaction is formed through the dual and at times conflicting forces of harmony and conflict. However, a significant overall limitation or difficulty with Schmalenbach's Bund model is found in the understanding that 'the search for community should be considered in most cases [. . .] as being a search for communion'. In light of the Nancean critique of community as communion discussed above, this formulation significantly undermines Hetherington's otherwise promising assertion (discussed earlier in this chapter) that the Bund concept has the potential to accommodate difference and otherness.

With respect to the second set of perspectives and the work of Hillery and Goode, this examination showed that certain understandings and definitions of community (while sometimes vague) are consistent and persist over time. Most notable is the ongoing importance of factors such as common ties and social interaction for definitions of community, regardless of whether they are locally based in particular geographical regions or areas, or geographically dispersed social aggregations, such as many professions and 'virtual community'. In the context of the earlier examination of virtual community, the work of Hillery and Goode is significant for three reasons. First, it situates virtual community within a broader history of community thought and points to an increasing convergence between computer-mediated understandings of community and more established understandings. Secondly, and as a result of the first, it implicates these computer-mediated forms of community in broader philosophical debates regarding community (such as those presented by Nancy and Derrida). Thirdly, and significantly for the work of the next chapter, Hillery's research found that 'area' (or local 'place') is not precluded from definitions of community but is consistently regarded as one of the less important elements in these definitions. In early writing on virtual community, these on-line social formations were celebrated for the way that they were seen to exist unfettered by the constraints of place and geography. The significance of Hillery's finding (and Goode's article) is that it qualifies the claims made about virtual community, and demonstrates that a reduced emphasis on place in definitions of community has been long established (and well-documented) for decades previous to interest in teletechnologies like CMC.

Finally, Nancy and Derrida argue that existing notions of community are inadequate, and both develop philosophically intricate alternatives in response. These responses raise multifaceted and difficult questions concerning the concept of community in general, and in respect to how the notion of community might be approached in relation to networked, computer-mediated communications. The merit of the approach taken by

Jean-Luc Nancy and Jacques Derrida in particular in rethinking community is found, once again, 'in the struggle with the tensions of singularity and commonality, of integration and difference' (Willson, 2006: 173). Moreover, their approach is commendable for recognising that this struggle is 'interminable and unresolvable' but should not be abandoned because of this fact (173). That is to say, it is in Nancy's and Derrida's (and to a somewhat lesser extent, Schmalenbach's earlier) individual attempts to think the risk of the limit of community—or to think the risk that takes place through exposure of and at the limit of community—that marks this work as significant. It is the work of Nancy and Derrida that holds greatest promise for thinking afresh not just the emergent phenomena of virtual community, but also the category of in-common community in general. However, one significant issue that still needs to be addressed in relation to their philosophical reorientations is the issue of how the subject and subjectivity is figured (or not) in these accounts. Thus, a fuller articulation of the overall promise and this specific shortcoming of the poststructuralist critique of and response to the problem of community will be withheld until the final chapter of this book.

Therefore, having examined the notion of community here, attention now turns to the idea of place and its many complexities: more specifically, where the previous chapter examined CMC and virtual community, and the present chapter enlarged on ideas of what 'community' might mean, the next chapter adds to this material by examining CMC and place. Together these analyses explore how the concept of 'place' relates to and engages with the notion of 'community'.

3 Haunting Affects
Place in Virtual Discourse

'It is important to remember that virtual community originates in, and must return to, the physical.'

—Allucquére Rosanne Stone (1992: 113)

'Metaphor is never innocent. It orients research and fixes results.'

—Jacques Derrida (2001a: 17)

'Milieu: *mille lieux*, a thousand places.'

—Edmond Jabès (1993: 15)

Making sense of the complex, three-way interactions of teletechnologies, place, and community constitutes the primary concern of this book. In Chapter 1, discussion focused on *community* and teletechnologies, where it was revealed that the notion of 'virtual community' has proven to be a particularly contentious term for describing the sorts of communicative practices that are performed via computer-mediated communications technologies. Chapter 2 extended this debate by suggesting that the contentious nature of the appellation of virtual community is a result of the complicated legacy bequeathed by the history of the word 'community', and that the debates concerning virtual community need to be situated within this wider history.

This chapter introduces the third key term or element in this project, 'place', and how the notion of place circulates in writing on teletechnologies (especially computer-mediated communication and virtual community). In so doing, it initiates the process of discovering its role and importance (if any) for thinking about and living in the digital age. Thus, this chapter has two primary aims. The first is to develop a fuller understanding of the notion of place, both as it is generally understood and as it pertains specifically to understandings of virtual community. A key question here is to ask what kind of conception of place is being invoked in writing on teletechnologies. The second, more polemical aim is to explore the prevailing view that place is marginal to considerations of computer-mediated communication exchange and virtual community, and whether ideas of place have any relevance in the context of teletechnology use. Do teletechnologies 'displace' place? Or, is place a more resilient and persistent notion? Does it have ongoing relevance for understanding teletechnology use and computer-mediated social interaction?

In pursuit of these aims, the analysis of this chapter draws from a wide range of disciplinary areas which concern themselves with ideas of place, including philosophy, geography, sociology, media and communications, architectural theory, and interdisciplinary studies of globalisation.

UNDERSTANDING PLACE

Dealing with the notion of place is difficult due to its lack of definitional clarity and precision. Even in as compendious a study as Edward Casey's *Getting Back into Place*, place is everywhere present but seemingly nowhere defined (Casey, 1993). Basic dictionary definitions do little to resolve general understanding of the term. For instance, *The Australian Concise Oxford Dictionary* offers thirteen variations, which range from broad references to space and its occupation, to the differentiation of types or 'sub-categories' of geographical space and the occupation of these spaces (including, in order of increasing expansion: a residence or dwelling; a group of houses in a town; a town square; a village, a town, a city; an area or region) (Hughes, Michell, and Ramson, 1992: 863; for more detailed discussion, see Casey, 1997). In practical terms this often means that place is taken to be linked to some kind of locality, with some sort of boundary. The problem with such definitions and with attempts at definition in general, as Edward Relph observes, is that 'place is not just the "where" of something; it is the location plus everything that occupies that location seen as an integrated and meaningful phenomenon' (Relph, 1986: 3; see also, Norberg-Schulz, 1976: 3). Lukermann (1964) describes this as involving complex integrations of nature and culture that have developed and continue to develop in particular locations, and which connect flows of people and goods to other places.

An even more expansive perspective on place is to suggest that difficulties of definition and experiential expression are in fact due to the notion of place being all-pervasive, structuring and shaping every facet of our lives and of our negotiation and experience of the lived world (Casey, 1993). In this respect, difficulties in grasping the notion of place are very much like the difficulties attending the category of the quotidian or everyday. As Maurice Blanchot (1987: 14) says of the everyday, 'whatever its other aspects, the everyday has this essential trait: it allows no hold. It escapes'. Its pervasiveness renders it as platitude. But, as Blanchot adds, 'this banality is also what is most important, if it brings us back to existence in its very spontaneity and as it is lived' (13). And so it is with place. The pervasiveness of place and its plurality of forms means that it allows no hold; its ubiquity and diffuseness is what makes place most important as it informs and shapes lived existence.[1] Thus, place can be understood as all-pervasive in the way that it informs and shapes everyday lived experience—including how it is filtered and experienced via the use of teletechnologies.

Other useful understandings of the term include the idea of place as being a structure of feeling or generated by 'fields of care' (Tuan, 1977), or more radically, as open and relational rather than bounded (Massey, 2005), or more fundamentally (but no less radically), as that which 'is integral to the very structure and possibility of experience' and 'within and with respect to which subjectivity is itself established' (Malpas, 1999: 32 & 35).

Some of these more nuanced understandings of place offer rich possibilities for understanding computer-mediated social interaction and will be taken up again in the final chapter. However, the more general—if rather limiting—understanding of place as linked to some kind of locality with some sort of boundary is certainly how place is commonly understood in discussions of networked communication.

In setting out on this examination of place and CMC, it is important to recognise that understandings and applications of place are often developed in complex and sometimes contradictory ways. Grasping something of this complexity and contradiction here is crucial, as it forms an important context or background to the subsequent focus on CMC and place.

Place has long been considered a problematical notion insofar as it is associated with strategies of exclusion and domination: 'the desire for some simple return to authentic local roots in "place" has been shown to be enmeshed in practices of cultural domination' (Dovey, 2002: 45; see also, Keith and Pile, 1996). These are the precise sentiments of Jean-Luc Nancy who, in his critique of community, as explored in the previous chapter, argues against an understanding of community as communion because this understanding suggests 'fusion into a body'—a monolithic form or identity—which denies difference and otherness and promotes exclusion (Nancy, 1991b: xxxviii). Nancy writes:

> The community that becomes *a single* thing (body, mind, fatherland, Leader . . .) necessarily loses the *in* of being-*in*-common. [. . .] (Nothing indicates more clearly what the logic of this being of togetherness can imply than the role of *Gemeinschaft*, of community, in Nazi ideology.) (Nancy, 1991b: xxxix)

Place is central to this critique of community, and separating the two is considered essential if community is to be reconceived in non-restrictive terms. As Van Den Abbeele remarks:

> What Nancy [. . .] deftly disconnects [. . .] is the assumed immanence of communal identities to demarcated geographical spaces in the form of towns, lands or nations. In its most vulgar formation, this relation

appears of course as the nationalist ideology of blood and soil. (Van Den Abbeele, 1997: 15)

Philosophical deliberations on community are by no means the only arena where place is 'disconnected'. There is, for instance, what Casey (1993: 10) terms the 'modernist myth that place can be discounted and set aside for the sake of space or time' (see also Casey, 1997: 131–193).

Mostly, however, place manifests itself and is understood in complex and often contradictory ways, as is illustrated in Manuel Castells's writing on globalisation. To cite one example, on the one hand, Castells writes:

> Localities become disembodied from their cultural, historical, geographic meaning, and reintegrated into functional networks, or into image collages, inducing a space of flows that substitutes for the space of places. (Castells, 1996: 375)[2]

On the other hand, at a later point, Castells makes the following qualification to his overall argument regarding 'networked society':

> The space of flows does not permeate down to the whole realm of human experience in the network society. [. . .] The overwhelming majority of people, in advanced and traditional societies alike, live in places, and so they perceive their space as place-based. A place is a locale whose form, function and meaning are self-contained within the boundaries of physical contiguity. (1996: 423)

Such contradiction and ambiguity is by no means isolated, and it is perhaps not surprising that the term 'globalisation' itself has become something of a conduit or pivot for the (re)consideration of global/local tensions in terms of articulating 'place' (Crane, Kawashima, and Kawasaki, 2002; Held and McGrew, 2002; Jameson and Miyoshi, 1998; Scholte, 2000; Waters, 1995). These debates argue, among other things, that 'the new industrial system is neither global nor local but "a new articulation of global and local dynamics"' (Amin and Robins quoted in Castells, 1996: 392).

It is in this context that 'glocalisation' has emerged as an important notion for capturing global-local tensions (Kraidy, 2001: 39; Robertson, 1995). The term was imported into social and cultural theory by Roland Robertson (1995) in order to correct what he saw as one of the 'major weaknesses' attending use of the word 'globalisation': that is, 'the tendency to cast the idea of globalization as inevitably in tension with the idea of localization' (40). This is problematic, he argues, as it denies the full extent to which 'the concept of globalization has involved the simultaneity and the interpenetration of what are conventionally called the global and the local' (30). Thus, for Robertson, the term 'glocalisation' serves a specific, rhetorical purpose: it works to emphasise the point that globalisation has involved and will continue to involve 'the creation and the incorporation of locality, processes which themselves largely shape, in turn, the compression of the world as a whole' (40).

What is clear from Robertson's arguments is that to engage with globalisation and the 'space of flows' (to use Castells's specific phrase) requires an ongoing concern for singularities and particularities of place and locality. In short: place, while in many respects a troublesome and contested term, it is nonetheless indispensable (Scriver, 2002: 4). To adapt the words of Edward Casey (1997: ix), we require places in which to exist; we are immersed in places and could not do without them.

These ideas about globalisation provide a useful context for the following discussion of CMC and place for two reasons. First, issues of CMC and place are situated firmly within globalisation debates and a broader 'global cultural economy' (Appadurai, 1990) that involves flows of people, media, technology, capital, and ideas, from, to, and across places. Second, this context provides a good example of how place, while imprecise in definition, is complex in understanding and application. This is also the case with CMC. The following discussion of CMC and place teases out something of this complexity and contradiction.

COMPUTER-MEDIATED COMMUNICATION AND PLACE

It could be argued that teletechnologies in general and CMC in particular both appear to contribute to a dislocation of place. As Joshua Meyrowitz argues in an important passage from his influential book, *No Sense of Place*:

> Changes in media in the past have always affected the relationship *among* places. They have affected the information that people *bring* to places and the information that people have *in* given places. But the relationship between place and social situation was still quite strong. Electronic media go one step further: They lead to a nearly total dissociation of physical place and social "place". When we communicate through telephone, radio, television, or computer, where we are physically no longer determines where and who we are socially. (Meyrowitz, 1985: 115)

Quite distinct from this is the severe view that, 'new [globalised] communications technologies [such as CMC] have been trumpeted as heralding the ultimate "death" of geography' (Morley, 2003: 439).

For example, with the emergence of cyberspace and virtual community it has been argued that place is not only dislocated, it is altogether abandoned, forming, according to the more extreme techno-boosterist accounts, a 'post-geographical' (Parrish, 2002), 'disembodied' realm that is free from the constraints of physics and the messy contingencies of the flesh (Spiller, 1998a). Such a position is no longer representative of prevailing attitudes to the Internet and CMC. In more moderate accounts of CMC, place is understood as a metaphor in the staging of computer-mediated social interaction, much like the way that the cinematic notion of *mise-en-scène* is understood. But even this metaphorical invocation of place is fiercely resisted at times (White, 2004). This resistance rehearses similar claims to exclusion discussed above in relation to community.

But CMC enjoys a far more ambiguous relationship with place than is perhaps suggested by the above formulations in that, at the same time recourse to the notion of place in the discourse of the virtual is actively resisted, it is also apparent that, in the context of CMC and virtual community, place persists in manifold ways, some examples of which are explained below.

To begin with, there is the banal ('realspace speaks up, whatever we may be doing online, whenever we get hungry' (Levinson, 2003: 39)) and the anecdotal (where the continuing pertinence of the question, 'Where are you?', in Internet chatroom conversations has been noted (Morley, 2003: 452)).

Place also persists metaphorically through the widespread use of architectural and urban planning motifs and metaphors in the graphical user interface design of many virtual communities, where users meet in virtual 'rooms' and 'agora'. Indeed, spatial metaphors profoundly influence and shape the way we think about on-line social interaction in spite of their increasing contestation. This is so important to the issue of understanding place within the literature on virtual community and CMC that it will be addressed at length later in this chapter.

But the persistence of place is even more fundamental than is suggested in the above examples. This can be illustrated by considering that apparent paragon of virtuality: global finance.

Saskia Sassen points out that global finance is very much a 'glocal' industry in that it simultaneously occupies the abstract 'cyberspace' of global telecommunications networks and local places: 'Such firms' activities are simultaneously partly deterritorialized and partly deeply territorialized; they span the globe, yet they are strategically concentrated in specific places' (Sassen, 2000: 224–225; see also, Clark, 2005). The 2007–2010 'global financial crisis' has driven this point home, with its repercussions felt in quite specific locations, from New York and London, to Dublin, Reykjavík, and Athens.

Sassen's formulation is a particularly useful way of understanding the notion of place in relation to CMC and virtual community. Like global finance, on-line social interaction and virtual community are simultaneously partly deterritorialised and partly deeply territorialised. This interaction occurs in numerous ways. At its most basic level, computer-mediated communication is only possible due to the infrastructure that supports it, such as fiberoptic cabling, electricity grids, satellite coverage, and so forth (Mitchell, 1995, 1999a). The sense of spatiality that is generated on-line also depends on an infrastructure of sorts, including the actual hard drives which store email and other data (see, Marshall, 2001).

Furthermore, while on-line social interactions occur within the abstract space of computers, these interactions are, for the most part, generated on and restricted to computers which are located in a place (even if temporarily so, in the case of mobile CMC technologies). A similar point is made elsewhere, with one critic noting that, 'Most people I know use one or two computers to access the Internet. These computers are sited, and the siting often affects the kind of Internet use engaged in, and perhaps the kind of "space" opened out' (Marshall, 2001: 82).[3]

Curiously, this rather fundamental fact of human-computer interaction is often overlooked, or considered of limited importance insofar as it is apparently 'effectively excluded from the shared context of CMC discussion [. . .] since it is not an object of common knowledge' (Rooksby, 2002: 105). However, experiential evidence contradicts this. For example, in the case of a predominately social listserv, such as Cybermind (Milne, 2010; 2007; Marshall, 2000), the issue of where one is—and all the geopolitical and other complexities that attend this location—emerges time and again. Such interactions usually concern 'infra-ordinary' or everyday experiences of place, such as posting links to images of a participant's home, domestic activities, and even their computer workstation. In addition to this domestic data, there is also more detailed engagement with the social and political contexts from which a participant writes.

The deterritorialisation-territorialisation dynamic is also apparent when considering the different forms of computer networks and the interplay between them. Broadly speaking, these networks range from (and bridge) very localised forms, such as company or organisational Intranet systems and other forms of local area networks, to wider area networks and global networks, such as the Internet and World Wide Web.

The key point here is that both are located to varying degrees in places. Thus, while local area networks (as their name suggests) are more commonly associated with and tied to 'local place', wider area networks are no less place-located or place-dependent. Wider networks, to use Sassen's words, 'span the globe, yet they are strategically concentrated in specific places' via the aforementioned underlying infrastructure, and via connection with more localised networks. This concentration also occurs through usage. As David Morley observes:

> The relative density of internet web connections per square kilometre in different geographical locations varies enormously, and access to these technologies (and to the 'connectivity' that they offer) depends very much on where you are located in both geographical and social space [. . .]. The distribution of these new technologies frequently mirrors established structures of power, and flows of internet traffic tend to follow the routes laid down by previous forms of communication. (Morley, 2003: 440)

In addition, many virtual communities had—if not still have—connections with local and regional places. For example, in its early days, the WELL (short for Whole Earth 'Lectronic Link)—which developed in part from the Whole Earth Catalogue and the American hippie movement—had strong ties to the San Francisco bay area (see Rheingold, 1994; Turner, 2006). A further example can be found in Digital City Amsterdam, which was created by a group of Amsterdam-based activists and artists. While the *raison d'être* of this 'virtual community' shifted significantly from the original vision of its creators (it became an electronic portal for and promotional tool of the City of Amsterdam government), it nevertheless remained firmly anchored to Amsterdam as place (Lovink, 2002; Aurigi, 2005; Jankowski, Van Selm and Hollander, 2001).

Lastly, in addition to a large body of work which emphasises the apparent capacity of teletechnologies such as CMC to facilitate communication and community independent of geographical place, there is a contrasting and expanding literature on the use of teletechnologies in connecting remote communities to one another (Uncapher, 1999; Howard, 1999), and in developing existing, geographically based communities—an area of research often referred to under the rubric of community (or urban) informatics (see Arnold, 2004; Day and Schuler, 2004; Doheny-Farina, 1996; Fathy, 1991; Foth, 2009; Gurstein, 2000; Hampton, 2002; Hampton and Wellman, 2000; Harrison and Stephen, 1999; Hopkins, et al., 2003; Huysman and Wulf, 2005; Keeble and Loader, 2001; Schmitz, 1997).

In sum, and for all the above reasons, within existing critical discourse on CMC and the virtual there is a strong and mounting challenge to the idea that 'the physical and virtual occupy opposite positions in a dichotomous relationship [where] cyberspace exists in a separate dimension from the physical and possesses radically different, almost divine characteristics' (Ward, 1998).

Place within CMC?

Clearly, place persists in manifold ways which strongly shape not only our engagement with CMC and virtual community, but also place in general. But what has not been addressed is whether place extends to the space *within* CMC and virtual community. To put this another way—restating a question John Perry Barlow (1999) has asked—'Is there a there in cyberspace'? Can it be argued that CMC and virtual community actually contain or consist of places?

For many commentators and virtual communitarians, the answer is unequivocally, 'Yes' (Bruckman, 1996; Grossman, 1997, Stone, 1992; Mitchell, 1995). But is this an accurate assertion? And what motivates this affirmative response? Emma Rooksby (2002: 129–130) suggests that, 'what is at stake in such claims can broadly be stated as "some forms of shared social life in CMC are as legitimate as any other form"'. But, according to Rooksby (2002), 'the truth of this claim can be preserved without reference to CMC place' (130). In other words, the idea that place can and does in fact exist within CMC and virtual community is for Rooksby a misnomer, a point she develops at length via reference to the states of sleep and death:

> We describe death as a parting or a going away, and talk of people going off to sleep. The places of sleep and death are places constituted imaginatively by reference to someone's absence from social engagement. Their constitution is simple, being nothing more, conceptually, than a non-engagement in social discourse. We may, of course, embroider on this, making it a *there* of some sort. But the basic move is the most important, since it signifies that we think of some non-spatial qualities of people as metaphorically place-related. In other words, we do sometimes talk of place without having in mind a place that is in *any way* spatially locatable. [. . .] The ascription of *there-ness* to those who are distracted, or asleep, are ways in which a merely *social* absence is figured as a *physical* inhabitation of another place. So, in using CMC we concentrate on it in a way that tends to make us socially absent from our immediate surroundings, and in touch with people in a way that is vivid and often immediate. (Rooksby, 2002: 132—original emphasis)

In other words, creating 'a sense that one is *in* some place that does not exist doesn't conjure up in any more than an imaginary way the place one senses' (132); rather, 'the sense in which CMC constitutes a place in which people can be or feel together will be only metaphorical' (126).

To summarise the preceding discussion, it has been argued that place persists in relation to human-computer interaction in its broadest sense in manifold ways. As for the issue of whether place exists (and persists) within CMC, it has been suggested that it does so metaphorically, but not in any literal sense. In other words, within the 'space' of CMC exchange, there would seem to be some sense of a 'soft where' but no internal or concrete 'there'. As Rooksby argues, 'the affect of place *haunts* moveable and remote institutions [like CMC] rather than constituting them' (2002: 128—emphasis added).

The precise wording of Rooksby's formulation is important, and the fuller implications of what it means for metaphors of place to 'haunt' computer-mediated social interaction needs to be explored further. A second, related concern will then be addressed: the suggestion that recourse to place-based metaphors actually leads to a denial of the multiplicity of space and place. This denial contributes to the persistent and troubling conception of the Internet (cyberspace) as distinct from 'real' or 'actual' space, as well as promoting exclusion and flattening difference. The discussion to follow below examines this engagement with place-based metaphors in the literature. This focus on metaphor should by no means be taken to suggest that place cannot be understood and experienced *simultaneously* in both a literal sense (as discussed above) and in a metaphorical sense. Rather, the argument that is developed here is that it is place in a metaphorical (rather than literal) sense that dominates much of the early phases of scholarship on the Internet and CMC.

THE RULE OF METAPHOR

The reduction of place to metaphor described above ignores the fact that a persistent and subtle force is exerted by metaphor in general. Metaphor is traditionally understood as a linguistic structure that implies similarities between two ostensibly dissimilar things. The 'essence of metaphor', as George Lakoff and Mark Johnson (1980: 5) explain, 'is understanding and experiencing one kind of thing in terms of another'—one of which is familiar, the other usually less so. For writers and readers, the power of metaphor rests in the fact that, 'The familiarity of the "known" domain can offer initial guidance to investigate and to plumb the "unknown" domain' (Fleckenstein, 1995: 111).

Herein rests the appeal of—and appeal to—metaphor in Internet discourse. It is the reason metaphors have been central to early imaginings of cyberspace (such as in Gibson's novel *Neuromancer*, which draws heavily on architectural metaphors in its representation of the 'space' of cyberspace), especially the many anecdotal accounts of cyberspace which 'resemble the old "travelers' tales"', accounts of adventurous trips from the civilized world to newly discovered, exotic realms' (Wellman and Gulia, 1999: 170). Metaphor proliferates in these accounts, a way of familiarising the 'unknown' via comparison with the 'known' (Milne, 2000). The relative unfamiliarity of cyberspace and computer-mediated communication is made comprehensible via comparison with more familiar notions and experiences, such as surfing, navigation, exploration, frontiering, settlement, transportation, highways, sites, the desk, office and home, and architecture and urban planning. Obviously, this list is by no means exhaustive. Here discussion is limited to three dominant types of metaphor: navigation and transportation metaphors; pioneer metaphors; and architectural metaphors. Each has proven popular in terms of use, yet each has also proven contentious and been viewed as restrictive.

Navigation and Transportation Metaphors

The prefix 'cyber' is derived from the Greek root *kybernan*, which means 'to steer or guide' (Tofts and McKeich, 1998: 19). In the literature on cyberspace, the most literal use of this root meaning is in the metaphor of navigation, where, 'this nautical figure is [considered] appropriate for a world imagined as a bit-stream of information flows' (19). The way that we 'navigate' these flows—the multiple pathways and hyperlinks of the Internet—and the reported addictive nature of this activity, often pursued as an end in itself, has also invited comparisons with 'surfing'—a metaphor which is said to be 'suggestive of [the Internet's] dynamism and hedonism' (Tofts, 1999: 23).

As for transportation metaphors, one of the more influential has been the description, from the early 1990s, of the Internet as an 'information superhighway'. Generally attributed to then US Vice President Al Gore, the information superhighway metaphor reached its peak in 1996, and has declined steadily in popularity since then (Blavin and Cohen, 2002: 269, note 22 & 271).

Pioneer Metaphors

Settlement metaphors and pioneer narratives proliferate in those texts which conceive of cyberspace not so much as pure information space but as a rich social space. One such text is Howard Rheingold's *Virtual Community*, which was originally released in North America in 1993 with the subtitle 'Homesteading on the Electronic Frontier' (this was subsequently altered in later additions to, among other things, 'Finding Connection in a Computerized World'). As discussed in the first chapter, Rheingold's text has been a key vehicle in the early promulgation of the community metaphor, a subset of the pioneer metaphor. The frontier as the promise of community is also evident in John Seabrook's *Deeper*:

> The landscape of the Net was not the great wide-open landscape of buffalo herds and antelope, although I had imagined it was in the first year of my travels. The frontier was more communal now. The frontier lay inside the group. (Seabrook, 1997: 131)

For Douglas Rushkoff (1994: 16), on the other hand, the metaphor's metaphysical possibilities are of greatest appeal; he conceives of cyberspace as 'the next dimensional home for consciousness', a 'timeless dimension', a 'boundless territory' known as 'Cyberia'. All three examples support Gregory Ulmer's (1994: 26–31) suggestion that the frontier metaphor dominates our current understandings of digital media.

Architectural Metaphors

The third, and most prolific metaphor type used to describe CMC is the architectural metaphor. Indeed, architectural metaphors proliferate to the extent that it is not possible to examine them in detail here. It will suffice to offer the following summary of their recurrence, reach, and ongoing relevance. In the context of cyberspace and human-computer interaction in general, architectural metaphors are employed in manifold ways. Within computer science, for example, the metaphor of architecture is widespread and explicit and serves an important organisational function, where 'the study of computer architecture is the study of the organisation and interconnection of components of computer systems' (Ibbett and Topham, 1989: 1). Architectural metaphors have also proven immensely popular in computer interface design (most notably, Microsoft 'Windows' and its associated 'desktop' iconography). And in the design of the graphical user interfaces of many virtual communities—where users inhabit 'rooms' or frequent various 'agora', 'cafés', 'bars', and other social spaces—the use of architectural and urban planning metaphors has been widely documented (see, for example, Mitchell, 2000; Morningstar and Farmer, 1992; Reid, 1995; Damer, 1998) (see Figure 3.1).

At an even more fundamental level, the architectural metaphor is said to lie at the heart of a wider history of ideas, structuring thinking in general and philosophical argument in particular. 'Philosophers since Plato', Kojin Karatani (1995: 5) writes, 'have returned over and again to architectural figures and metaphors as a way of grounding and stabilizing their

Figure 3.1   An example from the now defunct WorldsAway 'virtual community', which illustrates the prevalence of place-based metaphors in the interface design of virtual environments.
Source: William J. Mitchell (2000): 118. © MIT Press. Reproduced with permission.

otherwise unstable philosophical systems'. Karatani describes this return as, 'the will to architecture'—a drive, he argues, that characterises all of Western metaphysics, from early Greek philosophy to the present day (see also Brodsky Lacour, 1996, 1999; on the equally crucial influence of architectural metaphors in literature, see Cowling, 1998).

The Problem of Metaphor

The relative merits and limitations of these three recurrent metaphor types is not the aim of this discussion (see Adams, 1997). Rather, what is significant about them is that they are all, at core, species of *spatial* metaphors. This is important insofar as spatial metaphors are at the heart of many of the quarrels concerning cyberspace and CMC. Indeed, much of the anxiety, scepticism, and even hostility regarding virtual community tends, at some point, to circle back to a fundamental concern for the use and implications of spatial metaphor (see, for example, Robins, 1999).

Perhaps one important reason for this can be found in Derrida's assertion that 'metaphor is never innocent. It orients research and fixes results' (2001a: 17). Evidence for this is demonstrated clearly in the case of the three metaphor types already mentioned.

In the case of transportation, what is most troubling about this metaphor for some commentators is that it describes the Internet in strongly regulatory terms:

> Highways are built by the state to serve the automotive needs of the people, subject to state and (indirect) federal regulation. If the Internet is a highway, then government can regulate it for the safety of those who pass on it. (Blavin and Cohen, 2002: 269–270)

For those attempting to make sense of the social spaces of the Internet, this metaphor is of limited appeal. It 'connotes a transfer of information', not social interaction and habitation:

> Though cars use the highway to travel the distance between two destinations, no one 'lives' or 'resides' on the highway [. . .]. What is not intuitively obvious is that the information superhighway is itself an account of *space*, but *not place*. (Blavin and Cohen, 2002: 270—original emphasis)

In other words, the highway metaphor privileges one aspect of the medium only and not the many other possible uses and users of this medium:

> [The highway metaphor] focuses our attention on the road, the infrastructure, and away from the people and 'vehicles' that traverse it, away from the road-side, away from the interaction of road and place. (Jones, 2001: 54)

The concern is that the highway metaphor promotes a narcissistic inflation of the self at the expense of the 'other':

> We are led to believe we are in power, we are the ones 'surfing', or 'using', and others cannot see us [. . .]. The seeming absence of the other focuses away from economic and political issues, and directs us toward ourselves. (Jones, 2001: 54; see also, Pelikan Strau, 1997)

One response has been to propose an alternative, nautical metaphor which is considered to be more appropriate for capturing the nature and wider impacts of our engagement with the medium. This is the metaphor of a boat's wake: 'as we travel along an information "path", we leave behind a wake, though we may not leave behind tangible and permanent markers' (Jones, 2001: 54). In other words, when users interact via the Internet, they not only travel from place to place, they 'also create a "field" [or force] of influence and meaning around themselves' that radiates outwards like the wake of a passing boat (Jones, 2001: 54). However, when one recalls the meaning of the prefix 'cyber', and its implication of self-determination through 'steering', there is a suggestion that to supplant the highway metaphor with a nautical metaphor does not necessarily overcome or escape issues of self-interest, power, and the absence of the 'other'. This issue notwithstanding, what motivates a preference for one (the nautical) over the other (the transportational) is a desire to take regulatory considerations of Internet use and content and recast them in terms that more directly link the study of the Internet and communication to the study of social relations (55). Even then, however, as has been noted elsewhere, nautical or navigational metaphors do not 'fully evoke what is meant by cyberculture as a lived reality', especially insofar as they fail to capture the way that 'electronic networks stand in between the real world and our perception and experience of it' (Tofts and McKeich, 1998: 19).

In the case of the pioneer metaphor, the basic difficulty with the twin tropes of 'homesteading' and 'frontier' is twofold. On the one hand, it is argued that both feed a nostalgic, pastoralist myth of community—a myth, as discussed in the previous chapter, which is arguably threaded throughout the entire history of community thought (Bell, 2001: 98; Wellman and Gulia, 1999: 187). On the other hand, it is argued they perpetuate Western colonialist narratives, particularly of possession, oppression, and dispossession, as well of the imposition of private property conceptions upon cyberspace (Sardar, 1996; Hunter, 2003; Olson, 2005). As one source puts it, 'spatial metaphors become not merely useful tools for making the revolutionary changes of the information age less strange and unsettling, but a ready mechanism through which to manage and regulate the alien phenomenon' (Gumpert and Drucker, 1996: 32).

Architectural metaphors face similar challenges. Setting aside their wider philosophical application for the moment, which will be explored at length below, the use of architectural metaphors in interface design has been criticised for being non-intuitive, restrictive, and conservative (Thomas and Warren, 2000; Johnson, 1997: 61). Their usage in CMC research more generally is also resisted on the basis that it promotes exclusion and forecloses alternative ways of conceiving of the social space of CMC (White, 2004).

In sum, the above three examples—of transportational and navigational, pioneering, and architectural metaphors—all serve to illustrate the following points:

1. metaphor selection is never free of complication, and any comparison intended to clarify also has the capacity to obfuscate, and ill-chosen metaphors—even if they don't initially appear as such—can prove particularly detrimental (Johnson, 1997: 107);
2. insofar as metaphor is intertwined with rhetoric, metaphor is never innocent, and the rhetorical uses of metaphor always influence and shape the meanings that are generated by, and the meanings which accumulate around, a given metaphor;
3. metaphors orient research—setting the terms of reference of and agenda for research—and they fix results. The community metaphor is a prime example. More than just an efficacious symbolic term for characterizing virtual social relations, as Fernback (1999: 204) suggests, the tenacity of the community metaphor is such that it has dominated consideration of the social aspects of computer-mediated social interaction for the best part of two decades and at the expense of other, equally productive—not to mention more provocative—approaches (Wilbur, 2000: 55; Robins, 1999: 169–170).

Architectural Metaphor in the Text of Philosophy

The above three points are amplified further when one considers the role of spatial metaphor in the history of philosophy and, particularly, Jacques Derrida's sustained critical engagement with this history. In examining this engagement, I develop the argument that Derrida's critique of the use of spatial (especially architectural) metaphor in the text of philosophy parallels the use of spatial metaphor in writing on the Internet, CMC, and virtual community (which I hereafter refer to collectively as the discourse of the virtual). These parallels hold important implications for thinking about, and generating alternative ways of understanding, place metaphors and their relationship to CMC and virtual discourse. In order to tease out these parallels and their implications, it is first necessary to outline Derrida's arguments about metaphor and philosophy.

Architecture and architectural metaphor can be understood to be foundational to philosophy (or metaphysics, as Derrida prefers) insofar as the figure of architecture serves to 'ground' philosophical thought.[4] As Mark Wigley explains:

> The question of metaphysics has always been that of the ground on which things stand, even though it has only been explicitly formulated in these terms in the modern period inaugurated by Decartes. Metaphysics is no more than the attempt to locate the ground. Its history is that of a succession of different names (*logos, ratio, arche,* and so on) for the ground. (Wigley, 2002: 7)

The conventional relationship of architecture and metaphysics is very much one-sided, where 'philosophy treats its architectural motif as but a metaphor that can and should be discarded' (Wigley, 2002: 16). In other words, the 'marriage' of architecture and philosophy, as traditionally understood at least, is one of convenience, and a temporary one at that:

> [Philosophy] produces an architecture of grounded structure that it then uses for support, leaning on it, resting within it. The edifice is constructed to make theory possible, then subordinated as a metaphor to defer to some higher, nonmaterial truth. (2002: 16).

In order for this to work, and for philosophy to do its ('higher') work, architectural metaphor—which is also to say, representation and ornament—must be controlled, kept in check (59–95). In truth, however, metaphor and philosophy negotiate a far more 'complex and restless dynamic' (19).

It is precisely this rather too neat subordination of metaphor by (and in) philosophy that leads Derrida to remark on the orienting and fixing function of metaphor, and which makes him so suspicious of spatial and architectural metaphors. This issue of the subordination and ultimate dismissal of the architectural metaphor by philosophy is central to deconstruction and Derrida's (1997a) ongoing critical engagement with the Western metaphysical tradition (and his critique of presence, logocentrism, etc.).

Deconstructive discourse is indebted to Heidegger, and the term 'deconstruction' itself is a 'remobilisation' or 'translation' of Heidegger's *Destruktion* and *Abbau* (Wigley, 2002: 41). According to Derrida, '*Destruktion* is not a destruction but precisely a destructuring that dismantles the structural layers in the system'; *Abbau* 'has a similar meaning: to take apart an edifice in order to see how it is constituted or deconstituted' (Derrida, 1985: 86–87). As Wigley explains:

> In remobilising these terms, Derrida follows Heidegger's argument that this 'destructuring' or 'unbuilding' disturbs a tradition by inhabiting its structure in a way that exploits its metaphoric resources against itself, not to abandon the structure but to locate what it conceals. (2002: 42–43)

This process of 'unconcealing' reveals the extent to which philosophical structure depends and is built on the very flaws it seeks to disguise.

This is a crucial point. Deconstructive 'shaking' may well fracture the 'ground' on which philosophy is 'erected', revealing an abyss. But 'a structure does not simply collapse because it is erected on, and fractured by, an abyss' (Wigley, 2002: 42). Rather, abyss is in fact a 'structural necessity' of all structure (Derrida, 1997a: 163). Philosophical structure actually depends on and is built on the very flaws it seeks to disguise; indeed, its strength is located in these structural flaws (Wigley, 2002: 42–44). This insight will later prove helpful for thinking about the 'place' of place in virtual discourse.

Conceiving of structural flaws as a source of strength leads Derrida to pay close attention to the place and function of spatial metaphor (and architectural metaphor in particular) in the text of philosophy. From this close reading, Derrida argues that the traditional relationship of metaphor to discourse is in fact overturned. He writes, 'metaphor is less in the philosophical text (and in the rhetorical text coordinated with it) than the philosophical text is within metaphor' (Derrida, 1982b: 258). Or, as he puts it elsewhere, 'When the spatial model is hit upon, when it functions, critical reflection rests within it' (Derrida, 2001a: 17). Derrida's point, as Wigley (2002: 17) explains it, is that philosophy and the metaphysical tradition is 'orchestrated by what it thinks it employs'.[5] This provides important insight for thinking about place and the literature on ICTs.

It has been argued that Derrida resists the traditional treatment of spatial metaphor in philosophy by demonstrating how space is 'never a contingent element that can be simply bracketed out in favor of some higher immutable and immaterial constant' (Wigley, 2002: 70). This is clearest in relation to Derrida's understanding of language. As he writes in *Of Grammatology*, 'that language must traverse space, be obliged to be spaced, is not an accidental trait but the mark of its origin' (1997a: 323). Derrida's response is to 'unsettle' this tradition by rethinking space as 'spacing'. Like so many of the terms Derrida employs, 'spacing' has no fixed or singular meaning (Wigley, 2002: 73). Across its usage, however, 'spacing' is generally deployed as a generative concept:

> Spacing is a concept which also, but not exclusively, carries the meaning of a productive, positive, generative force. Like *dissemination*, like *difference* it carries along with it a *genetic* motif: it is not only interval, the space constituted between two things (which is the usual sense of spacing), but also spac*ing*, the operation, or in any event, the movement of setting aside. (Derrida, 1981: 106, note 42).[6]

There is much more that could be said on this point, particularly concerning the directions that Derrida develops this argument and idea of 'spacing' in his own writing and other projects (see, for example, Derrida, 1995a, and Kipnis and Leeser, 1987). However, it will suffice here to state that Derrida's general point about spatial metaphor in the text of philosophy also holds for place metaphors in virtual discourse: place cannot be brack-eted out from, or made marginal to, the discourse of the virtual. In the present context, then, this too offers important insight for thinking about place and the discourse of the virtual.

The Haunting Affect of Place in the Discourse of the Virtual

Derrida's critique of metaphor offers fresh and productive ways for think-ing about metaphors (especially the metaphor of place) in the literature on cyberspace and information and communications technologies. This occurs in at least two main, interconnected ways.

The first is that the 'convoluted economy' and 'curious strategic role' of architectural metaphor in the text of philosophy is similar to the way that metaphors of place circulate in and structure the discourse of the virtual. It was argued above that philosophy 'produces an architecture of grounded structure that it then uses for support' which is then 'sub-ordinated as a metaphor [in order] to defer to some higher, nonmaterial truth' (Wigley, 2002: 16). The idea of place serves as the paradigm of ground as support, but is then 'subordinated' as a metaphor in order to defer to some higher 'truth' or condition which maintains that cyber-space is 'ungrounded'. In other words, place is used metaphorically as a structural framework for communication and then largely dispensed with. As Rooksby (2002: 129–130) points out, 'place' is often invoked in discussions of the social life of CMC in order to legitimate it as a valid form of social interaction and 'community'—after which, its function is essentially fulfilled. One possible explanation for this can be found in the observation that, 'an architectural metaphor [in this context] serves less to define a sense of place than to indicate a set of social functions' (Adams, 1997: 159). What is more, these social functions are enabled by electronic communications which, as noted earlier in this chapter, are said to operate on the understanding that 'where we are physically no longer determines where and who we are socially' (Meyrowitz, 1985: 115). Thus, to argue that place cannot exist in CMC except as a meta-phor gives support to the idea of CMC and 'cyberspace' (the Internet) as 'dematerialised', 'ungrounded'. This, however, veils or denies cyber-space's ongoing 'emplacement'.

Furthermore, this rather instrumental view of 'place', which claims that it can remain only a metaphor, is, to use Jeff Malpas's words, 'simply a reassertion of a particular and fairly narrow view of the nature of place—a reassertion that seems to ill-accord with the complex character of the con-cept' (1999: 38). For instance, while it is possible—indeed preferable—to discuss place in relation to CMC in *both* its metaphorical and literal sense, this is rarely done in any meaningful way in the literature under examina-tion here.

The second point is interconnected with and extends the first. It draws on Derrida's argument that philosophical structure actually depends on and is built on the very flaws it seeks to disguise. Wigley (2002: 92) takes this even further by proposing that, 'architecture, which appears to be the good tame ornament of philosophy, is actually the possibility of philosophy'. A similar logic could be applied to the role of place metaphors in virtual discourse (especially writing on CMC): that is, place, which appears to be the good tame ornament of the virtual, is actually the possibility of the virtual. This means that the way in which place is used as a metaphor in writing on CMC and the virtual can be read, alternatively, as invoking the poststructuralist argument regarding the function of a suppressed term in an apparent oppositional pairing.

Perhaps the best known example of this argument concerning binary pairings can be found in Derrida's (2001c) critique of the work of anthropologist Claude Lévi-Strauss, especially the latter's methodological approach of 'bricolage' and its associated figure of the 'bricoleur'. Lévi-Strauss (1966) conceives of the bricoleur as an inventive anthropological investigator who assembled a meaningful cosmology from the random elements encountered in day-to-day life. The bricoleur is set in opposition to the 'engineer' who is said to be everything the bricoleur is not: 'rational', 'civilised', uninventive, inflexible, and dull. In other words, the two notions—bricoleur and engineer—are framed according to a strict binary opposition. This has been referred to as the logic of 'totalizing negation', whereby 'each of these binary oppositions produces the myth of a privileged term by producing the myth of an absolutely opposing term' (Lucy, 1997: 99). It *is* a myth, because with each apparent 'oppositional' pairing, not only does one term permit the very possibility of the other term, but this possibility comes about because each term *contains* the other, contains elements of this difference. As Derrida writes of Lévi-Strauss's two figures:

> As soon as we cease to believe in such an engineer and in a discourse which breaks with the received historical discourse, as soon as we admit that every finite discourse is bound by a certain *bricolage* and that the engineer and the scientist are also species of *bricoleurs*, then the very idea of *bricolage* is menaced and the difference in which it took on its meaning breaks down. (Derrida, 2001c: 360–361)

In other words, 'the engineer is a myth produced by the *bricoleur*' (360). To translate this to the present context, the argument here is that the historical framing of the virtual as 'unbounded' and 'dematerialised' is a myth produced by the insistence on place as only metaphorical. Moreover, these metaphorical constructions are based on quite limited underlying conceptions of place which deny the full complexity of this concept.

This has important implications for understanding metaphors of place within virtual discourse. Emma Rooksby argues that place as such, 'does not *constitute* remote institutions' like virtual community (2002: 128—emphasis added). Nevertheless, to conceive of metaphorical uses of place in the above ways is to acknowledge, at the very least, that 'place' (in its broadest sense) exerts considerable *constitutive influence* on virtual discourse in general and on virtual community in particular.

To put the matter in spectral terms, place *haunts* virtual discourse and virtual community. Any attempt to 'exorcise' place from virtual discourse—especially by some commentators and techno-boosterists in the 1980s and 1990s—only results in temporary 'displacement' prior to its reassertion as an ongoing *haunting affect of place* (Rooksby, 2002: 128).

This notion of the spectrality (Derrida, 1994), or haunting affect, of place is critical to how the notion of place is understood in this study, specifically in relation to the literature on CMC technologies. The persistence of place and the way it operates as the structural figure—the 'weakness'—that structures virtual discourse means that place continues to 'haunt' or shadow telecommunications and (what Derrida terms) the whole apparatus of teletechnology in general (Derrida and Stiegler, 2002).

One further, possible partial explanation for the persistence of place as an idea in relation to CMC exchange, and especially in the interface design of virtual communities, is that it seems to connect with growing interest in and concern for issues of 'emotion' and 'other modes of feeling' in connection with place, particularly within geography (Thien, 2005; Davidson and Milligan, 2004; Thrift, 2004; Anderson and Smith, 2001). Such concerns have long been of interest to CMC researchers. For example, many (especially early) 'talk-and-text' studies of CMC often focus on the reliance on textual cues and associated icons—such as the appropriately named 'emoticons'—by CMC users to register emotion and feeling (for discussion, see Hine, 2002). This interest in the textual expression of emotion and feeling also often dovetails with broader concerns, such as questions of (dis)embodiment (see Chapter 1). Historically, however, communicative considerations generally took precedence over place considerations. This is why geographical interest in emotion and feeling is especially interesting in the context of the present discussion. Of particular relevance is Davidson and Milligan's (2004) contention that, 'the articulation of emotion is [. . .] spatially mediated in a manner that is not simply metaphorical', and that 'our attempts to understand emotion or make sense of space, are, thus, somewhat circular in nature' (523–524). They propose that 'emotions are understandable—"sensible"—only in the context of particular places. Likewise, place must be *felt* to make sense' (Davidson and Milligan, 2004: 524—original emphasis). This understanding accords with Jeff Malpas's (1999: 32) claim that, place 'is integral to the very structure and possibility of experience'. Moreover, both these perspectives—that by Davidson and Milligan and that by Malpas—have the potential to reorient and significantly enrich existing understandings of the emotional or 'affective' dimensions of on-line social interaction.

The haunting affect of place is also significant insofar as metaphor in general is all-pervasive. Malpas's earlier criticism of metaphor notwithstanding, metaphor remains important and cannot be ignored: it spatialises language, structures the way we think, and, as Derrida has argued, is part-and-parcel of the whole history of Western metaphysics (with the discourse of the virtual situated firmly within this history). As Derrida puts it elsewhere, through the use of a double negative, 'There is nothing that does not happen with metaphor and by metaphor. Any statement concerning anything that happens, metaphor included, will be produced *not without* metaphor' (Derrida, 1998: 103).

This offers a further means by which Derrida's critique of metaphor and textuality can assist: it presents productive ways of approaching the issue of how to proceed with a critique of spatial metaphor in virtual discourse. In large part this is achieved simply by recalling Derrida's famous formulation from *Of Grammatology*: 'there is no outside-text (*il n'y a pas de hors-texte*)' (1997a: 158). What Derrida means by this (among other things) is that 'our reading must be intrinsic and remain within the text' (159). Commenting on the impossibility of thinking about the 'structurality of structure' from outside Western metaphysics, Derrida writes:

> There is no sense in doing without the concept of metaphysics in order to shake metaphysics. We have no language—no syntax and no lexicon—which is foreign to this history; we can pronounce not a single destructive proposition which has not already had to slip into the form, the logic, and the implicit postulations of precisely what it seeks to contest. (2001c: 354)

The same is true of spatial metaphor. In the present context, what is required is a certain ethics of reading (as Derrida might say), a commitment to close reading, which is attentive to the 'spatialising' of language and/through metaphor. Such a reading interrogates metaphor and language, and asks certain questions of the 'spatialising' affect of metaphor:

> How is this history of metaphor possible? Does the fact that language can determine things only by spatializing them suffice to explain that, in return, language must spatialize itself as soon as it designates and reflects upon itself? (Derrida, 2001a: 17–18)[7]

Derrida (2001a: 18) concludes this passage by adding, 'this question can be asked in general about all language and metaphors. But here it takes on a particular urgency'. And so it does in the present context, in our consideration of the 'place' of place within virtual discourse.

Within this discourse, there are valid reasons for being suspicious of spatial metaphor. At the same time, as Donna Haraway would say, there is room for taking '*pleasure* in the confusion of boundaries, and for *responsibility* in their construction' (1991: 150—original emphasis).

But it is equally important to recollect that it is only possible to critique spatial metaphor from within, by working with and against these metaphors, recognising their limits and working at these limits. It is also important to acknowledge the vitality of these metaphors (with 'vitality' intended here in a double sense, suggesting *importance to* and *animation of* virtual discourse) and their ongoing relevance to understanding teletechnologies and the techno-social.

CYBERSPACES AND OTHER PLACES

The ideas and understandings that have emerged from this detailed consideration of metaphor—about the persistence of place, its ongoing importance as a wider concept, and its inextricable context for the virtual—raise a number of further issues. Key among these is Emma Rooksby's (2002: 131) argument that recourse to place-based metaphors can actually lead to a denial of space and 'placial' multiplicity *and* of singularities of place. Smith and Katz (1996: 75) argue a similar point, when they suggest that the problem with spatial metaphors is that they presume that space is not problematic. They write:

> The problem lies not with spatial metaphors as such, but with metaphors that depend on a very specific representation of space: *absolute space.* [. . .] Absolute space is the space that is broadly taken for granted in Western societies—our naïvely assumed sense of space as emptiness—but it is only one of many ways in which space can be conceptualized. (Smith and Katz, 1996: 75)[8]

This appeal of (and to) the idea of 'absolute space' is also associated with the idea of space as abstract. For example, 'space' is often considered more abstract than 'place': 'What begins as undifferentiated space becomes place as we get to know it better and endow it with value' (Tuan, 1977: 6). These moves are particularly strong in writing on cyberspace and other forms of CMC, where 'space' is often privileged over 'place' on the basis that space is considered 'more abstract' than place, and therefore most befitting of a realm that so often in the past has been described as 'dematerialised'. According to Doreen Massey, however, this is not necessarily (or not always, and certainly not strictly) the case. 'In an age of "globalization"', Massey writes, space 'is no more than the sum of all our relations and connections (friendly and antagonistic) and, like place, it too is continually being made' (2002: 24–25). According to Massey, 'to not see space in this way is to understand it as "abstract"' (25). Massey's ideas are significant for thinking about the relations between teletechnologies and place, and will be revisited in the final chapter.

The above point emphasises the social construction of space and of place, their 'production' through individual action, social interaction, and political and economic processes (Lefebvre, 2000; Harvey, 1990). The issue is that these complexities concerning the production of space and place are far from adequately accounted for in the literature on CMC. This is a problem, because, as Lefebvre (2000: 41) argues, the 'perceived-conceived-lived triad (in spatial terms: spatial practice, representations of space, representational space) loses its force if it is treated as an abstract "model"', which it so often has been in the populist and boosterist accounts of cyberspace and CMC from the 1980s and 1990s.

The dependence on absolute space, and a consequent lack of attention to the proliferation, particularities and production of spaces and places, also underpins the way cyberspace has traditionally been understood. According to Lance Strate (1999: 383), a common misconception is the conception of cyberspace as singular when in fact it might be more appropriate to conceive of it as plural (see also Adams, 1998; Adams and Warf, 1997).

A key benefit of conceiving of cyberspace as multiple, Strate (1999) explains, is that it 'guards against the dangers of high-level abstractions removed from reality many times over, in favor of a concrete and grounded emphasis on difference' (383). It provides, in other words, a starting point from which to 'survey the varieties of cyberspace' (383). This is particularly important given the proliferation in meaning and application of the term cyberspace since the 1980s and 1990s, which can make the concept 'problematic from a scholarly perspective' (382).

What is needed in response, Strate argues, is a 'more precise linguistic map', 'a survey and taxonomy of the cyberspaces' (383), which he then proceeds to set out in the form of a tripartite taxonomy, with each order encompassing several different spatial forms. His delineation of the various forms of cyberspace draws out a number of important points. For example, and as already noted, discussing cyberspace in the plural ensures a 'grounded emphasis on difference' (383). Indeed, what becomes apparent through this taxonomy is that, in the literature on cyberspace, not all forms of this space are granted equal coverage—some are clearly favoured over others. A key example of this is the privileging of what he terms 'conceptual cyberspace' over that of 'physical cyberspace':

> The one variety [of cyberspace] that is often overlooked is physical cyberspace, the material base of computers, monitors, disk drives, modems, wires, etc., and their users. Most discussions instead focus on

conceptual cyberspace, the sense of space generated within the mind as we interact with computer technology. (Strate, 1999: 384)

Such a partisan approach has a number of consequences. One consequence of overlooking the 'physical cyberspace' of computers and their users is a kind of 'aporia' or amnesia, where it is sometimes forgotten that, 'contributors to CMC discussions are spread out over a multitude of places, which they inhabit in a bodily sense' (Rooksby, 2002: 131). Thus:

> To overlook this proliferation of places (and contexts) from which CMCs are sent is to risk ignoring important differences [. . .] such as the different fields of action open to different persons using CMC, or the range of shared actions possible for members of a CMC community, given their lack of propinquity. (Rooksby, 2002: 131)

This oversight of the physical also has wider, sensorial implications. Two aspects said to be constitutive of our experience of cyberspace are the notions of 'immersion' and 'presence'. Strate places both under the heading 'perceptual cyberspace'. Thus, to privilege 'perceptual cyberspace' is also to privilege experiences which create a sense of immersion and presence. But, as Strate points out, 'perceptual cyberspace is just one of the varieties of cyberspace', not the default form. By extension, then, while it might be desired by some, the sense of immersion and presence generated through CMC use constitutes *an* experience, not *the default* experience, of 'cyberspace'.

A further implication of privileging 'perceptual cyberspace' over 'physical cyberspace' is that 'cyberspace is almost always seen as non-physical space' (Strate, 1999: 390). For example, cyberspace has been described as a space 'parallel' to geographical space (Tække, 2002), and of invoking 'a tantalizing abstraction [. . .] of disembodied immersion in a "space" that has no coordinates in actual space' (Tofts and McKeich, 1998: 15). However, as Strate (1999: 389) points out, 'a physical basis for all forms of cyberspace can be established that is consistent with the physical basis of our everyday life', from the structure of the computer (its chips and circuits), to the actual location where information is magnetically stored, as well as the whole electronics and telecommunications infrastructure needed to connect computers to power sources and to each other (Strate, 1999: 390; Marshall, 2001: 87–90).

And perhaps the major consequence—itself a result of the preceding consequences—is the common, lingering, and decidedly troubling dichotomous conception of 'cyberspace' as a realm distinct from 'actual' (or 'Cartesian geographical') space. Although increasingly challenged, this is a view that was once widely circulated by techno-boosterists (Spiller, 1998a) and sceptics alike (Boyer, 1996), and which constituted a key basis (or point of departure) for many examinations of cyberspace. Thinking through this dichotomy in order to think beyond it is a necessary and vital task. This need also points to one of the key strengths of Strate's taxonomic study: the idea of cyberspatial multiplicity enabling broader spatial critique, which in turn loops back to inform understanding of 'the cyberspaces'. Strate (1999: 389) argues that by dismantling the cyberspace/physical space binary opposition, we are able to achieve a better understanding of the social construction of space in general. The reverse also holds true: 'we can further our understanding of cyberspace by considering the basis of our conceptions of physical space' (389).

'UNDOING-PRESERVING' PLACE

A key argument of this chapter has been that in much of the writing on CMC from the mid-1980s to the mid-1990s denies the all-pervasive impact of place, the full extent to which place persists. The persistence of place occurs on many levels and in a variety of contexts. These include: the recurrent reference to place and geography in text-based exchanges within virtual community; the place-associated design of many virtual community interfaces; the use of CMC technologies in community-building projects; the 'place' of everyday computer-use, such as the office or domestic home; as well as the notion of 'glocalisation' and the broader context of global-local interchange. Place also persists through the language and metaphors which structure the discourse of the virtual. Indeed, this discourse is thoroughly spatial and 'placial'.

To recognise the persistence of place is to take a significant step towards abolishing the major and long-held misconception that cyberspace exists independent of 'actual' space. It might well be possible to conceive of 'cyberspace as a space parallel to geographical space', as Jesper Tække (2002) does—although this denies the multidimensionality and plurality of cyberspace, of cyberspaces. However, it is difficult to support the assertion that 'new' globalised communications technologies realise the 'death' of geography. In fact, the notions of the virtual and of virtual community are thoroughly inscribed in geographical, spatial, and place-based terms. And it is precisely this philosophical 'ground', these terms of reference, which require sustained critical engagement. There is value in being suspicious of the dominance of metaphors of space and place and how these operate to 'orient research' and 'fix results'. Nevertheless, a definite balancing act, a tension, is implicit in proceeding from this scepticism. What is entailed is the difficult project that Spivak (1997: xliii) refers to as 'undoing-preserving': undoing-preserving binary oppositions, as well as the metaphorical language which circumscribes and shapes discourse. The task is to recognise and work with the all-pervasiveness of place (in both its metaphorical and more general sense), but also to work against and *up against* (at the limits of) this all-pervasiveness—to recognise, in other words, that as critics, we cannot step outside the text.

In addition to its all-pervasiveness, place emerges as a thoroughly techno-social concern—despite ongoing attempts to separate the techno-social from place in CMC literature. It is important to be attentive not only to the particularities of the multiple places that structure everyday CMC use, but also to a wider techno-social *milieu*—which, as Edmond Jabès (1993: 15) reminds us, can be read as *mille lieux*, a thousand places.

The next chapter continues to examine the notions of place and community, albeit as they are addressed *outside* of the literature on CMC. There the focus will shift from an examination of virtual community and 'cyberplace' to an examination of place and community within the architectural computing literature. Turning to architectural engagement with computing is valuable in order to examine how a more conscious engagement with a specific teletechnology, such as the computer, influences architectural understandings of place and even of community. The first of these chapters initiates this investigation by considering early architectural engagement with computing technologies.

4 Machines of Tomorrow Past
Early Experiments in Architectural Computing

> 'Those who mistrust the machine and those who glorify it show the same incapacity to utilize it. Machine work and mass production offer unheard-of possibilities for creation, and those who are able to place those possibilities at the service of a daring imagination will be my creators of tomorrow.'
>
> —Constant Nieuwenhuys (quoted in Wigley, 1998: 27)

In the introduction to her book *Always Already New*, Lisa Gitelman argues that media constitute 'very particular sites for very particular, importantly social as well as historically and culturally specific experiences of meaning' (2008: 8). Given this understanding, Gitelman positions the case study as a key tool for making sense of these various layers and experiences of meaning. Taking a similar stance, this chapter examines one particular phase of development (within Anglo-American and European contexts) where early possibilities of a particular teletechnology—the computer—were explored and tested. Specifically, the focus here is on architecture, a discipline that historically has been concerned with issues of place and community, and its early engagement with computing technologies.

In the following pages, the literature on architectural engagement with early computing technologies is examined within the period spanning the 1960s and 1970s, in order to explore what this explicit engagement might reveal about architectural understandings of place and community. This particular time-span proved to be a seminal and a boom period in the development and uptake of computing technologies within architecture (Reynolds, 1980: 2). As comprehensive accounts of early architectural computing are available elsewhere (Campion, 1968), the approach here is to consider four sets of representative texts that provide a contextual snapshot or cross-section of how computers were being used and written about within architecture and urban design at this time. This is followed by a specific case-study so as to provide a closer focus on one particular designer—the English architect Cedric Price—who was noted for his early experimental architectural engagement with computing and other teletechnologies.

The following questions are asked of this material. What design-related possibilities did computing technologies present for architecture at that time, and how were these possibilities taken up and discussed within the architecture literature at that time? What were the social ambitions of this earlier architectural engagement with computing? Furthermore, what was the broader context of this uptake, and what were the possible influences and forces attending how architectural computing was framed and received? Finally, and more generally, how might this early architectural engagement with computing contribute to the subsequent history of cyber-culture and technology use?

EARLY ARCHITECTURAL COMPUTING: FOUR TEXTUAL STUDIES

While computers were first developed in the 1940s and 1950s, by the early 1960s the wider possibilities and potential applications of computing technologies were already beginning to be realised. For example, in 1963 Ivan Sutherland (1963, 2003) unveiled his SKETCHPAD system which included a television-style monitor allowing the user to draw in lines of light on the screen with a special pen:

> Arcs, circles and straight lines could be produced with little effort, and repeated drawing elements or symbols could be defined once and there-after called up and positioned as many times as necessary. (Reynolds, 1980: 1)

The architectural possibilities of such a system quickly began to be realised. Two years after Sutherland, William Newman developed a similar, experimental system specifically for architectural application which stored standard building elements (such as walls, windows, floor slabs, and so forth), which the user could then select from and combine to create a plan on the screen (Reynolds, 1980: 1; Newman, 1966; see also, Fair, et al., 1966).

Developments in architectural computing snowballed over the ensuing decades. Important texts from this boom period include Serge Chermayeff and Christopher Alexander's *Community and Privacy* (1963), Yona Friedman's *Toward a Scientific Architecture* (1975), Nicholas Negroponte's *The Architecture Machine* (1970) and *Soft Architecture Machines* (1975c), and Peter Cook's, *Experimental Architecture* (1970).

Each of the above texts were written by established figures (and published by reputable publishers) within the academic community of architects and urban designers. They distil key trends and new ideas circulating within architecture and urban design at this time with respect to the possibilities and possible implications of computing technologies. These particular texts also provide vital clues as to the wider contributing influences and discourses that shape the reception and use of teletechnologies, such as the computer, within a field such as architecture.

Chermayeff and Alexander: Calculating Order and 'Community'

Serge Chermayeff was a Chechen born, British raised architect who later immigrated to the United States, while Christopher Alexander was a Viennese born British architect who, in 1958, also immigrated to the United States to teach at the University of California, Berkeley. Together they collaborated on the 1963 book, *Community and Privacy*. In the present context, this book is significant insofar as it represents a comparatively early example of explicit architectural engagement with the possibilities presented by computing technologies. In it, Chermayeff and Alexander are concerned with issues similar to those that were to preoccupy writers such as Jane Jacobs (1972) and Lewis Mumford (1979) a decade or two later: problems of sprawl, car dependence, and the over-industrialisation and unsympathetic development of cities. Chermayeff and Alexander write:

> Designed environments will be successful only if they respond to the most crucial pressures of our time. This means that they must resolve the problems created by often useless mobility, the ceaseless sounds and noises of communication and machinery, and the dissolution of the tranquillity and independence known in earlier cultures. (Chermayeff and Alexander, 1963: 105)

At the heart of the problems facing urbanism, they argue, are the dual and at times competing needs for community and privacy, and the issue of how to better integrate the two: 'the anatomy of the urban structure as a whole and [. . .] the anatomy of dwellings' (149). A key starting point in responding to these challenges, they claim, is to rethink 'the dwelling form': the home.

> The dwelling form must react to the pressures of modern communications media. Yet [. . .] the housing designed for the new mobile look-listen society is obsolete. [. . . W]hile supplying the standard, mechanized forms of comforts and conveniences, [it] ignores completely the need for variety in day-to-day life. (1963: 117)

This essentially amounts to a thinly veiled attack on what they term the 'all-too-open plan' of the streamlined modernist home. Their belief is that the 'electronic age' demanded smaller, partitioned spaces (127–128).

In response, they develop a list of thirty-three 'basic requirements' that are 'called for by the need for privacy' and which respond to the aforementioned pressures (154). However, the challenge they face is how to narrow this unwieldy list of items into something more manageable for generating practicable design solutions—'this is the crucial question in any design process, for countless different views of the problem are possible' (159). For the two authors, this is where the computer comes in. 'Problems of this kind', they write, 'cannot be solved without the help of electronic computers' (160).

> The computer, while unable to invent, can explore relations very quickly and systematically, according to prescribed rules. It functions as a natural extension of man's analytical ability. (160)

Chermayeff and Alexander are at pains to stress that the computer is merely an aid or a tool for use by the designer: 'the machine is distinctly complementary to and not a substitute for man's creative talent' (160). Elsewhere, Alexander reiterates his belief in the primary problem-solving function of the computer, adding the following caution:

> Until we have thought [. . .] problems through so far at the conceptual level that we encounter unanswerable complexities in them, and until we have managed to describe these complexities so precisely that an army of clerks could help us unravel them, there is no sense in trying to use a computer. (Alexander, 1965: 6)

As Alexander elaborates, 'distortion and triviality [. . .] is bound to occur whenever people *try to apply* the computer to design, rather than waiting until they *have* to use the computer because they are confronted by a complexity which they cannot resolve without it' (1965: 7–8).

This cautionary perspective notwithstanding, Chermayeff and Alexander see the use of a computer as warranted and valuable. While little is said about the actual input and computation process itself, the authors state that they used an IBM 704 computer at Massachusetts Institute of Technology (MIT), supplied with 'appropriate instructions', to make the necessary calculations (1963: 161). 'In a few minutes', they write, 'resolution of the apparent chaos of thirty-three items into groupings is made clear' (161).

Chermayeff and Alexander's study is somewhat problematic in the context of this study. To begin with, they provide no clear definition of community. In addition, this vision of urban renewal, based on the centrality of the domestic home (see 215 & 254), might be seen by some as both nostalgic and therefore conservative and potentially limiting. These issues notwithstanding, the key point here is to understand how computing technologies are conceived and utilised in their project. This technology is clearly understood and employed solely as a computational and problem-solving device—a tool in the development of 'conditions for restoring genuine variety' in residential and urban design (117), and is not employed as a representational device. What is clear from their work is that, for Chermayeff and Alexander, computing technologies are a design aid only and do not make the designer obsolete or any less central to the design process.

Friedman and the 'Flatwriter'

Yet, for Hungarian born French architect and urban planner Yona Friedman (writing a decade later than Chermayeff and Alexander, in 1975), reconsidering the place of the designer is necessary. For Chermayeff and Alexander (1963: 111), 'the biggest obstacle to improved design standards is the obsolescence of designers themselves'. In contrast, Friedman sees the democratisation of the design process as an imperative:

> Since 1957, I have been working on a theory which would free the client from the 'patronage' of the architect, and at the same time, I have been looking for a way to make the architect useful to the client. (Friedman, 1975: xi)

As Friedman sees it, a growing proliferation of clients and their individual needs 'implies that the whole process of which the architect is a part must be changed' (3). The change that Friedman advocates involves a 'scientific', rule-based system of design—one that generates a palette of design choices to meet and serve the specific needs of future users: 'In our scheme, instead of an architect, the future user encounters a *repertoire of all the possible arrangements (solutions) that his way of life may require*' (8—original emphasis). This approach does not seek to eliminate the architect. Rather, it recasts the role of the architect as one who 'constructs the repertoire' (9), whose job it is to 'establish the repertoire, the instructions, and the infrastructure' (11; cf Manning, 1973: 50).

Friedman develops an 'application of the repertoire' in his proposal for a computational design machine called the Flatwriter (see Figure 4.1).

Friedman's vision for this machine is that it will contain a 'repertoire of several million possible plans for apartments' (53). Each future inhabitant of a city accesses this information by responding to a series of questions, in order to choose the plan, characteristics, and location of their desired apartment. Friedman refers to this as the first feedback loop (7–9). The Flatwriter machine then calculates and informs them and other existing city inhabitants of the consequences of these decisions and each future decision (59–60). This forms the second feedback loop (7–11). According to Friedman,

> The Flatwriter thus puts a new informational process between the future user and the object he will use; it allows for almost limitless individual choice and an immediate opportunity to correct errors without the intervention of professional intermediaries. (60)

More than this, Friedman also believes that the Flatwriter facilitates a 'more democratic social organization' (140).

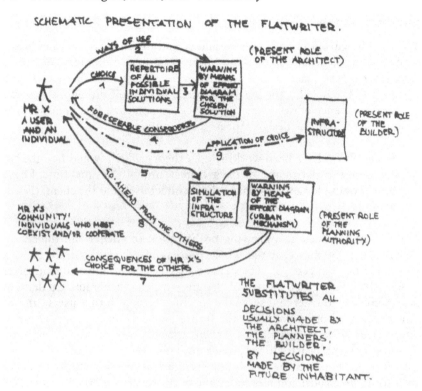

Figure 4.1 An 'organigram' or schematic diagram of Yona Friedman's proposed 'Flatwriter' architectural computing machine.
Source: Yona Friedman (1975: 54). © MIT Press. Reproduced with permission.

Negroponte's 'Architecture Machine'

There are similarities between Friedman's thinking and the ideas expressed around the same time on the same topic by Greek-American architect, and co-founder of MIT's Media Lab, Nicholas Negroponte in his two books, *The Architecture Machine* (1970) and *Soft Architecture Machines* (1975c).

For instance, Negroponte, who met Friedman as a graduate student in 1964 and was to later work alongside him, shares Friedman's concern 'for removing the architect as middleman between a user's needs and their resolution in the built environment' (Negroponte, 1975a: ix). Also, like Friedman, Negroponte sees the architect's role as that of interpreting physical form, rather than determining the conditions for this form. He also holds to Friedman's belief that computing technologies provide the potential means to achieve this marriage of user requirements and built form (Negroponte, 1975b). As Negroponte writes at the beginning of *Soft Architecture Machines*:

There is no doubt that computers can help in the humdrum activities of making architecture tick: smooth circulation, sound structures, viable financing. But I am not interested in that—I am interested in the rather singular goal of making the built environment responsive to me and to you, individually, a right I consider as important as the right to good education. (Negroponte, 1975c: x)

However, while Friedman is only able to imagine such a universal, plan-generating machine and the possibilities it would enable, Negroponte actually sets about developing the software necessary to realise such a machine (see Figure 4.2).

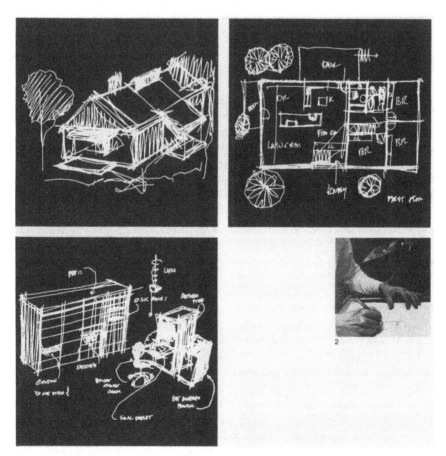

Figure 4.2 Examples of computer-generated architectural sketches produced on Negroponte and MIT's 'architecture machine'. While visual representation was clearly possible with this program, it was nevertheless resisted by Negroponte, who believed it obscured some deeper issues.
Source: Nicholas Negroponte (1975: 63). © MIT Press. Reproduced with permission.

According to Negroponte (1970), any architecture machine must go beyond providing feedback on the basis of a set of predetermined inputs (see Cross, 1977: 42–43). Rather, it is imperative, Negroponte (1970: 7) argues, for 'machines not only to problem-solve but also to problem-worry'.

Negroponte's vision for such an architecture machine goes well beyond that proposed by Friedman (and Chermayeff and Alexander for that matter). Negroponte envisages a computer-generated architectural 'cybernetic system'. Any computational system for architecture, Negroponte argues, has to be open to external stimuli:

> An adaptable machine must receive direct sensory information from the real world. It must see, hear, and read, and it must take walks in the garden. Information should pass into the machine through observation channels that are direct rather than undergo the mutations of transfer from the real world to designer's sensors to designer's brain to designer's effectors to machine's sensors. (Negroponte, 1970: 7 & 9)

Such a machine should also be responsive, not just to environmental stimuli but the issues that might emerge in response to these: 'An architecture machine that could observe existing environments in the real world and design behaviours [. . .] would furnish the architect with both unsolicited knowledge and unsolicited problems' (1970: 121).

Whilst acknowledging that 'no [fully] adaptable machine exists today', especially one which is able to deal with nuances of meaning and context, Negroponte nevertheless calls on architects and computer scientists to continue the task of building systems that can 'learn, can grope, and can fumble, machines that will be architectural partners, architecture machines' (121). As he writes in *Soft Architecture Machines*, 'I strongly believe that it is very important to play with these ideas scientifically and explore applications of machine intelligence that totter between being unimaginably oppressive and unbelievably exciting' (1975c: 1).

Negroponte's interest in the future possibilities of computer technologies hints at a key point of divergence between his work and that of Friedman. Whereas Friedman is interested in the use of computer technologies for the realisation of more democratic or egalitarian forms of social organisation, Negroponte leans towards the creation of responsive environments, and of the computer itself as part of such an environment. This vision for architectural computing is expressed in the third of what he calls the three 'potentials of the computer'. These are: '(1) the computer as designer, (2) the computer as a partner to the novice with a self-interest, and (3) the computer as a physical environment that knows me' (1975c: ix).

Nevertheless, the tension that Negroponte seeks to maintain in his work, and in his vision for an architecture machine, is a responsive, 'interactive', and ultimately 'immersive' technology which frees the client-user from reliance on the architect-designer, yet which remains fully committed to serving (and servicing) this client-user. As Sean Wellesley-Miller (1975: 129) writes in a short essay contained within *Soft Architecture Machines*, 'we should not forget that a building's final context of response is the needs and senses of its inhabitants'.

Cook's Survey of Experimental Architecture

A somewhat different, and perhaps more diffuse, picture of architectural computing emerges in the writing of the British architect Peter Cook, especially his 1970 book *Experimental Architecture*. Cook was one of the founding members of the influential Archigram group in the 1960s; at the time of publication of *Experimental Architecture* he had become Director of the Institute of Contemporary Arts in London, a position he held from 1970 to 1972.

Experimental Architecture is not a treatise on computing and its implications for architectural design. Rather, as the title declares, it is a survey text of late 1960s architectural experimentation and its implications. Concerned with documenting techniques and technological innovations that facilitate and contribute to experimentation in architecture, the text explores a variety of explicitly 'technologised' experiments in spatial construction, habitation, and community-formation, by the likes of Cedric Price, Archigram, Team 10, Arata Isozaki, and many others (see also, Zuk and Clark, 1970).

A key point of departure for Cook's survey is, as he describes it, the 'turn to eclecticism' in the experimental architecture of the 1960s:

> The word [eclecticism] itself had derogatory implications only a decade ago (architecture was to be pure and discriminating), but it now implies a positive openness and absorption of anything that might be useful to a project. (Cook, 1970: 14)

This 'bricoleur' approach to design is most strongly felt in the period's rapid up-take of new building materials (due to advances in materials manufacturing technologies), and—given the wide-reaching influence of the work of Canadian media theorist Marshall McLuhan (Wigley, 2006)—in the engagement with and 'absorption' of telecommunications and media technologies.

Thus, 'technology' is meant here in its broadest sense—from the structural and engineering innovations that permit the construction of space-frame, pneumatic, and geodesic structures, to the sorts of technologies that are the specific concern here: computing and communications technologies and their role in architectural experimentation of the period. Cook argues that, when combined, the 'opportunity of the material' (1970: 55) and the turn 'towards technology' act as a 'great force for a new architecture' (30).

What distinguishes Cook's text from those already examined in this chapter, is that computing technologies are not discussed as a representational device, but that teletechnologies—particularly computing and communications technologies—feature in *Experimental Architecture* as components of the architectural schemes themselves. In other words, teletechnologies constitute important tools within the experimental architect's overall suite or toolkit of available technologies. And it is the consistency in the overall selection and combination of these technological tools, Cook argues, which defines the 'problems' which experimental architecture is attempting to resolve.

Cook's thesis, in effect, is that 'experiment in architecture is the disintegration of architecture'—at least, that is, the disintegration of capital 'a' 'Architecture', which, Cook suggests, is more the result of a broader architectural dissatisfaction. He writes, 'I cannot resist the disclosure of an increasing feeling of dissatisfaction with the *role*, the *constraints* and the *formal mythology* of most architecture as such' (7). This feeling is based on 'an increasing awareness of the environment' and a 'decreasing respect for the relevance of institutionalized blocks of building and decrepit technologies propped-up by an elitist aesthetic language' (7).

He goes on to argue that there is an emerging architecture that is at once ephemeral and responsive. 'It is now possible,' Cook writes, 'to discuss architecture not only as throw-away but as more closely related to a particular situation at a particular point in time. It can be much more related to the ambiguity of life. It can be throw-away or addictive; it can be ad-hoc; it can be more allied to the personality and personal situation of the people who may use it' (67). It is in this context that he discusses such schemes as Price's 'Potteries Thinkbelt' and 'Fun Palace' projects (to be detailed later), as well as the work of others, such as Archigram, Dennis Crompton, Yona Friedman, R. Buckminster Fuller, David Greene, Arata Isozaki, Jean Prouvé, Team 10, the French Utopie Group, and more.

Given this position, it is perhaps not surprising that the 'opportunity of the material' and the 'turn towards technology' dovetail in two recurrent themes in Cook's study (see Figure 4.3). These themes are an abiding interest in mobility (and neo-nomadism), and challenges to the function and operation of the domestic home, within experimental architecture of this period. Both constitute key elements in his study. (Given their significance, both will be discussed further in Chapter 6—which addresses questions of teletechnologies, the domestic home, and mobility—where they are of greatest immediate relevance.)

Threaded throughout, and emerging from, this survey study, Cook develops the following further arguments.

First, experimental architectural 'forms the components of a see-saw activity between the rational investigation of techniques at one end and the enforced accentuation of the new and always extending limits of what might be termed "environment" at the other' (152). Technological innovation, mobility, and a critique of the domestic home, are key aspects of this see-sawing activity.

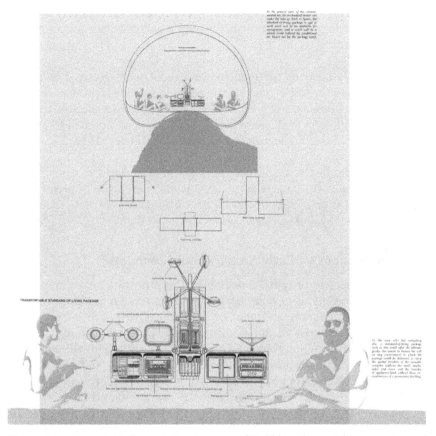

Figure 4.3 François Dallegret's 'Environmental Bubble' scheme. This image is used by Cook to illustrate experimental architecture's engagement in the 1960s with teletechnologies and questions of mobility.
Source: © 1965 François Dallegret, *Un-house. Transportable Standard-of-Living Package. The Environment Bubble.* Drawing No 5 of 6 produced for Reyner Banham's article, "A Home Is Not a House", *Art in America*, April 1965, China ink on translucent film 65,5x50cm and gelatine on clear acetate 76x76cm, Collection François Dallegret, Montréal. Reproduced with permission.

Secondly, these investigations can also be seen as part-and-parcel of a 'basic dialogue between movement, structure, and the possible transfer of events and their location within the structure' (101). 'The sift of these three conversations,' he writes, 'can be passed across most experimental projects' (101). This is an important point, and will be returned to in Chapter 6 in relation to contemporary mobile teletechnologies.

Thirdly, and fundamentally, running throughout Cook's study is a belief that technological innovation offers constructive possibilities for architecture. The general thrust of Cook's argument concerning the embrace of technologies—both material and communicative/

computational—by the experimental architects he surveys can be summarised according to Constant Nieuwenhuys's statement for a creative use of machines:

> Those who mistrust the machine and those who glorify it show the same incapacity to utilize it. Machine work and mass production offer unheard-of possibilities for creation, and those who are able to place those possibilities at the service of a daring imagination will be my creators of tomorrow. (Nieuwenhuys, quoted in Wigley, 1998: 27)

In the concluding passage to *Experimental Architecture*, Cook (1970: 152) writes that, 'If I have tried to suggest that the future of architecture lies in the explosion of Architecture, I have also tried to express something very optimistic' by and through this 'explosion'.

Initial Reflections on Early Architectural Computing

The first three sets of texts above show the potential for computers in architecture in generating complex computations and in problem-solving, rather than as a representational or form-generating technology. Chermayeff and Alexander's emphasis on the architect maintaining creative control would presumably preclude their consideration of the representational possibilities offered by computing. Friedman and Negroponte actively resist the potential of this technology for representation. For example, Friedman (1975: xi) warns about 'wandering off into visualisations'; Negroponte (1975c: 57) concurs, arguing that the 'picture-making part of computer graphics has obscured some deeper issues'. Even so, as Sandra Kaji-O'Grady (2004) observes, while 'they were not interested in the invention of a new [visual] aesthetic [they], nevertheless, were unable to avoid one'. Kaji-O'Grady notes that Negroponte's books include numerous black and white photographs 'documenting the equipment used in experiments in computer-aided vision and intelligence'. Meanwhile, Friedman's text contains over 80 figures, including 'flow charts and bubble diagrams, lists, examples of notation systems, graphs, matrices, tables, geometrical figures, rudimentary maps, and diagrams of networks, systems, structural types and architectural enclosures' (Kaji-O'Grady, 2004). In the same way, in *Community and Privacy*, Chermayeff and Alexander transform the fruits of their computational analysis into a whole series of floor plans. Thus, while the representational possibilities of computers are nominally eschewed for problem-solving objectives, representation remains inescapable in terms of outcome.

Nevertheless, and despite this seemingly inexorable pull towards representation, what remains of key interest in the present context is the way all authors understand the computer as first-and-foremost—indeed, almost exclusively—a problem-solving technology. This is notwithstanding Cook's more diffuse use, where these technologies form part of an overall scheme or program which takes architectural fixity, 'environment', and other factors, as its 'problem', to be solved.

Secondly, this problem-solving is underpinned by and anchored to a firmly 'humanist' agenda in the sense that architecture and architectural computing has as its primary aim the improvement of the human condition and the conditions of human habitation. The first two texts, by Chermayeff and Alexander and Friedman, both take as their primary point of focus the 'future user' of the contemporary city. In Negroponte, meanwhile, the 'humanist' agenda becomes somewhat harder to discern, but nevertheless he is adamant that what drives his interest in an architectural machine is a pursuit of 'the rather singular goal of making the built environment responsive to me and to you' (1975c: x). For this reason, he argues, architectural machines must 'problem-worry' as well as 'problem-solve' (1970: 7). In the fourth example, Cook continues the loosely 'humanist' agenda of the other writers, albeit in a more expansive sense, in that he is attempting to document those schemes that are most appropriate to the circumstances and social context of the period. Despite this interest in human habitation and well-being, however, there is very little discussion of or concern for notions of 'community'—even in the first of the four examples, where it is a key term in the title to Chermayeff and Alexander's text.

The third observation is that, in many respects, and perhaps unintentionally, place appears to be of reduced importance in comparison to other considerations. Across all four sets of examples, the dual emphasis on standardisation and individual choice leads to a retreat from questions of place, which is interesting given that the notion of *genius loci*, or 'spirit of place', was given such strong emphasis in the contemporaneous work of the architectural theorist Christian Norberg-Schulz (1976, 1980). As Cook (1970: 14) writes in relation to experimental architecture, 'now a generation of architects is growing up that is less interested in the local architecture of any region than in the pursuit of a certain attitude, from whatever part of the world it stems'. The irony, Cook argues, is that on the one hand, an underlying consideration of all experimental architecture is a consideration of 'questions of *place*, facility, equipment and the idiosyncracies of the users' (1970: 122—emphasis added). On the other hand and as already discussed, in the works gathered together in his survey there is a strong emphasis on mobility.

The fourth initial observation relates to systems theory and cybernetics and how—most notably in the first three sets of texts—these form key influences for architectural computing. This is understandable given that the systems approach and cybernetics enjoyed considerable support in architectural discourse during this period (see Handler, 1970; Ferguson, 1975).

Architectural interest in the systems approach has its origins in (among other sources) British and American research efforts during WWII and the Cold War era (Hughes and Hughes, 2000a, 2000b; Keller, 2002). It also drew heavily on the pioneering work of biologist Ludwig von Bertalanffy, who is generally credited with developing General System Theory (Davidson, 1983). Early English-language accounts of his model were published around 1950 (Von Bertalanffy, 1950, 1951). By the 1960s, knowledge of this approach was disseminated widely in a number of well-publicised books, including *Problems of Life* (1960), *Robots, Men and Minds* (1967), and *General System Theory* (1968). Von Bertalanffy's basic thesis is that 'modern technology and society have become so complex that traditional ways and means are not sufficient any more but approaches of a holistic or systems, and generalist or inter-disciplinary nature became necessary' (1968: xx). Von Bertalanffy proposes 'General System Theory' as a 'logico-mathematical field, the subject matter of which is the formulation and deduction of those principles which are valid for "systems" in general' (1950: 139). He argues that 'there are principles which apply to systems in general', whatever the nature of their component elements or the relations or "forces" between them', and that his General System Theory is therefore 'applicable to all sciences concerned with systems' (139).

Meanwhile, architectural interest in cybernetics drew from manifold sources, but particularly the equally pioneering and influential work of Norbert Weiner (as well as Alan Turing, Stafford Beer, among others). As Gordon Pask explains in *An Approach to Cybernetics*:

> Cybernetics is an interdisciplinary study of a physical assembly—it takes characteristics common to each discipline and abstracts them into a system to yield effective control procedures, efficient predictions, and acceptable unifying theories. (Pask, 1960: 17; see also, Lobsinger, 2000b: 138, note 63; Pask, 1969)

There would seem to be many similarities shared by the two theoretical approaches, systems theory and cybernetics, particularly in their joint emphasis on accounting for complexities within an overarching framework. For von Bertalanffy, however, to see them as synonymous is erroneous. Rather, one (cybernetics) is subsumed within the other (General System Theory):

> Cybernetics, as the theory of control mechanisms in technology and nature and founded on the concepts of information and feedback, is but a part of a general theory of systems: cybernetic systems are a special case, however important, of systems showing self-regulation. (Von Bertalanffy, 1968: 17; 1967: 65–69; 1951)

Despite these taxonomic border disputes, the architectural possibilities of both theoretical approaches are apparent and were quick to be taken up. For instance, as Handler observes:

Because the totality with which architecture is concerned is not one un-differentiated whole, it can be grasped only by understanding its parts, and, because these are numerous, by rationally ordering them in relation to each other and to the system as a whole. (Handler, 1970: 4)

Handler goes on to argue that, 'To see architecture as a system is to see the architectural process in an objective framework' (22). Such an approach is said to arm 'the problem solver with an understanding of how to organize the parts of a problem and how to derive consistent solutions' (25).

These influences are taken up in different ways in the architectural texts under examination here. For example, Chermayeff and Alexander freely quote from von Bertalanffy's *Problems of Life* in the many epigraphs that introduce each chapter of *Community and Privacy*. However, for Friedman and Negroponte, cybernetics takes a far more central role. For instance, Friedman's vision for his Flatwriter machine is that it will operate as a kind of 'cybernetic' system, involving information input, feedback, and control. Negroponte aims to extend this process even further in his development of an architecture machine. 'Computer-aided design', he writes, 'cannot occur without machine intelligence—and would be dangerous without it' (Negroponte, 1970: 1). Negroponte's vision of architectural computing owes a huge debt to the literature on computing, communications, cybernetics, and artificial intelligence—the bibliography to *The Architecture Machine* references many of the pioneers of all these fields, including J. C. R. Licklider and Arnold von Neumann (networked computing), Marvin Minsky and Alan Turing (AI), Marshall McLuhan (communications), and Norbert Weiner (cybernetics). More than this, it could be said that his understanding of architectural computing is inseparable from and incomprehensible outside of these theoretical influences.

Also interesting in the context of the present discussion is that these theoretical approaches came to be regarded in some quarters as synonymous with architectural computing (perhaps it is significant, in this respect, that von Bertalanffy describes systems theory as a 'logico-mathetical field'). This is not, by-and-large, regarded as a constructive union, however. Critics of systems theory condemn architects who embrace it for combining 'the needless jargon of the analyst' with 'unintelligible diagrams and interminable computer print-outs' (Ferguson, 1975: 3). Elsewhere, systems theory is characterised as 'synonymous with technocratic bureaucracy', and as simply 'a matter of building software' ('Editorial', 1976: 267; see also Rabeneck, 1976). Meanwhile, Jackson (2002: 70) argues that, 'the methodologies of systems analysis [and cybernetics] may be subject to a wide range of social and political interpretations', which might make their usage as a universal system somewhat problematic.

Despite these criticisms, in the context of the present discussion, there are further points to emphasise about this marriage of architectural computing and General System Theory and cybernetics. For instance, this relationship is illuminating insofar as it shows a wider theoretical and conceptual context in which these technologies were being deployed at that particular period in time. Furthermore, it shows that the issues facing architects were taken to be complex and increasingly unresolvable without equally sophisticated conceptual and practical tools, such as General System Theory and cybernetics in the first instance, and technologies such as the computer in the second.

The fifth and final observation makes a more general point. This concerns the way in which the computer is used to add vital overall support to the respective projects of the authors examined. This support is not just instrumental; it is also symbolic. Despite their various differences of opinion about the role of the design professional, these authors are interested in developing systematic and logical solutions to design problems. In this context, the computer lends significant symbolic support to their respective enterprises. In each of their texts, the computer is framed (albeit largely implicitly) as a 'scientific' technology, and in this sense it is both an instrument of and instrumental to their systematic—'scientific'—approach to architecture and urban design.

GENERATING FUN WITHIN TECHNOLOGISED SPACE: THE ARCHITECTURE OF CEDRIC PRICE

The above observations or findings form a useful basis from which to compare and contrast the material that follows below. In the remainder of this chapter, architectural interest in computing, and the theoretical traditions of systems theory and cybernetics, are developed further by examining in detail the work of one particular 'experimental architect' of the period: the iconoclastic British architect and theorist, Cedric Price—an architect, as noted earlier, who is significant in the present context for his early engagement with computing and other teletechnologies not as design tools but as integral components of his designs and overall architectural philosophy.

In this case, as well as looking at his writing, attention will also be paid to his designs because the two are intertwined. Three of his projects will be canvassed here: his 'Potteries Thinkbelt', 'Generator', and 'Fun Palace' projects. In particular, discussion is focused on the last of these: his unbuilt London Fun Palace Project, and the comparatively early incorporation into this project of computing and other communications technologies. Price did this as part of an ongoing critique of the conventions of architectural form. More specifically, he saw these technologies as one way of testing non-authoritarian ideas of use (multiplicity of use, and undreamt of use), encouraging chance encounters and, it was hoped, chance 'community'.

Critical to understanding his work, and the Fun Palace and its significance in particular, it is necessary to provide a picture of Cedric Price and his rather unique design philosophy.

Cedric Price was born in Staffordshire, England, in 1934 and died in London in 2003. Price was a larger-than-life figure, renowned for his acerbic wit and for being somewhat anti-establishment; he described himself as an 'anti-architect' and seemed to court controversy. For example, in one radio interview Price was asked what he would do about the medieval landmark York Minster. His reply: 'Flatten it' (Allford, 1984: 7).

He also despised the idea of built heritage and particularly heritage conservation. 'What is objectionable,' he once wrote, 'is the staggering conceit and arrogance of those who determine just what part of our built environment should be deemed sacrosanct.' (Price, 2003: 37) In a regular magazine column aimed at his fellow architects, Price once provocatively suggested: 'the best technical advice may be that rather than build a house, your client should leave his wife' (1966a: 483).

These are more than just throw-away one-liners designed to provoke. While they *are* designed to elicit a response, underlying all of them are three of Price's key interests: a love of paradox, a scorn of dogma, and a desire to improve the human condition (Allford, 1984: 7).

The above remarks also reveal much about Price's design philosophy and his understanding of the role of architecture and the architect. For instance, at the heart of his comment about the client and his wife, Price is posing what is something of a heretical question to his fellow architects: is a new building what is actually most needed in every situation?

And underpinning his antipathy to built heritage and his York Minster remark is a firm belief that a structure should last only as long as it is socially relevant. Once this relevance has diminished, he argues, the building should be demolished and replaced with something of greater immediate social relevance. Price refers to this as the principle of 'calculated uncertainty': 'the creation of temporary, adaptable structures that can be altered, transformed or demolished, serving the need of the moment' ('Cedric Price', 2003). Price's own 'Inter-Action Centre' in Kentish Town, England, is a case in point. In the initial scheme for the building, Price produced a client's manual explaining how to disassemble the building and recycle its components—one critic describes this document as 'something like a euthanasia guide and donor card rolled into one' (Baillieu, 1998: 7).

Price maintained that the role of architecture and architects is to solve problems, encourage choice and delight, and to develop ideas and possibilities rather than specific design solutions. These are the core tenets of what has been termed his 'philosophy of enabling', in which technology played a central part (Landau, 1984: 11).

Information and communications technologies such as the computer do not feature as instruments in the generation of Price's schemes at all. Rather, as in Cook's survey study of experimental architecture, these technologies, where they are present, are a part of the fabric of these schemes.

Throughout his career, Price maintained a very particular attitude towards technology and its use. As Royston Landau explains, technology had to meet certain criteria:

> First, there needs to be an appropriateness (perhaps to be found in electronics, but equally possibly in a primitive log cabin), and appropriateness will seldom be synonymous with the conventional. Secondly, technology will be used to play a *critical* role, meaning that it will be expected to take part in the architectural debate, perhaps through contribution, disputation or the ability to shock. And thirdly, technology must be placed in a particular and real context from which a framework of limiting constraints can be derived. (Landau, 1984: 11)

It is according to each of the above three senses that technology is employed in Price's projects, but particularly in the first of them to be discussed here: his unimplemented 'Potteries Thinkbelt' (or PTb) project (1964–1967). In this project, Price (1966b) used technology to think through issues of place and how a particular site could be adapted for use as a higher educational facility. The site was a decaying industrial belt-shaped site in north Staffordshire famed for its former pottery manufacture. It was chosen deliberately by Price due to what he saw as its desperate need for institutional and economic revitalisation. The proposal he developed provided for learning facilities, student accommodation, and a transport system that can easily connect the site to neighbouring centres and other regions (see Figure 4.4). Adhering to his principle of the 'non-plan' (Price, 1984: 38), Price described the 'real set of priorities' for the site as 'a firm re-assessment of housing requirements together with an avoidance in the first stage of development of any *civic* design' (1966b: 484). In addressing these housing requirements, the scheme included an array of possible accommodation configurations which would take best advantage of an existing and difficult topography (characterised by slag heaps and subsidence-prone soil). The centre-piece of the scheme was an ingenious use of the existing infrastructure of a largely abandoned rail system which, due to a large number of switches and stops, was unsuited to integration into the national rail network. For Price, this apparent liability emerged as an asset. The PTb facility, Royston Landau (1984: 13) explains, 'would house predominantly departments of science and technology, fast-changing subjects whose scope, size and life-span would be impossible to determine except in the short term.' This emphasis on the temporal and on temporal contingency opened a range of possibilities for rail-borne facilities:

> The fast-changing industrial engineering departments requiring cranes and gantries for replacing equipment were obvious candidates but Price

was also to propose teaching units in a large variety of different mobile enclosures which included combinations of inflatable lecture theatres, fold-out decks, library and information carrels and units and a series of capsule facilities, all with capabilities for combination and permutation and all available to be demounted and transferred as required. (Landau, 1984: 13)

The primary import of these possibilities for Price is that his scheme not only takes 'full advantage of the present means of individual mobility', but 'equally, it is so designed as to prevent its form and organization becoming restrictive' (Price, 1966b: 484). The local rail infrastructure also enables connection to national or international transport networks via what Price terms the 'Meir Transfer Area', a road-PTb rail-air link which provides facilities for rapid movement of people, goods, and hardware in and out of the PTb network (1996b: 485 & 487).

Even though unimplemented, Price's proposal represents a clever response to the ongoing call for urban revitalisation and non-automobile dependent transport planning. The scheme also carries significant implications for pedagogical architecture. According to Landau (1984: 13), the merit of this study is, 'not as an example of how railway carriages can be used for teaching, but as one of most powerful question marks ever placed against the architecture of university education'.

Figure 4.4 Montage developed by Cedric Price to illustrate his 'Potteries Thinkbelt' (PTb) proposal.
Source: Cedric Price, architect, 1934—2003. Perspective sketch of transfer area, Potteries Thinkbelt, 1964. DR1995:0280:621, 1966 gelatin silver print of photomontage (épreuve argentique à la gélatine d'un montage photo). sheet: 24,4 x 47,5 cm (9 5/8 x 18 11/16 in.) DR2006:0019
© Cedric Price fonds Collection Centre Canadien d'Architecture/Canadian Centre for Architecture, Montréal. Reproduced with permission.

In the context of the present study, one of the more striking aspects of the PTb proposal lies in its engagement with technology and questions of place, facility, and equipment. Price's study, to borrow a phrase from Marc Augé, pays close attention to *factors of singularity*—especially singularities of place and economies of place use. Moreover, as Mary Lou Lobsinger (2000a: 26) observes, 'the reuse of outdated channels of physical mobility— the network of the industrial revolution—is a convenience that facilitates the growth of a new mobility at the dawn of the information age'.

Crucial to this project, and characteristic of Price's general engagement with technology, is a twofold insistence. The first concerns what Landau (1984: 11) terms Price's 'philosophy of enabling': where buildings and technologies must act as catalysts which facilitate and encourage social and spatial interaction for and between users (Spiller, 2002: 84). The second concerns what Price calls 'free-space' (1984: 54) and an insistence on the notion of 'calculated uncertainty': 'the creation of temporary, adaptable structures that can be altered, transformed or demolished, serving the need of the moment' ('Cedric Price', 2003). As Price says of PTb's housing units, 'they would be expandable and, of course, expendable. No one would be straight-jacketed into a fixed community' (Price, 1966b: 483). Peter Cook (1970: 67) refers to this as architectural experiment's 'strategy for future change'.

In Price's 'Generator' project (1976) a decade later, 'calculated uncertainty' was to become a property, or function, of the building itself. The Generator was the name Price gave to a proposed building for the Gilman Paper Corporation in Florida ('A Building', 1981: 743). The building was to be comprised of 150 cubes, all of which could be assembled in a variety of different configurations (see Figure 4.5):

> Spaces and enclosures would be created, using orthogonal and diagonal geometries, with walls, screens, gangways, and the volumes would be fully serviced by systems including air conditioning as well as communication channels. (Landau, 1984: 15)

Connected to a central computer program which was to control a mobile crane, the entire Generator structure could be rearranged at will to create new configurations according to different use requirements (15). What is more, if the structure was not reorganised by its users for some time, the computer, it was said, would register 'boredom'; this would result in proposals for 'unsolicited change' (15; 'A Building', 1981: 743). While the proposal was ultimately abandoned, the Generator project is considered by some to represent one of the first 'intelligent' buildings, and an important symbiosis of Price's thesis of 'enabling' and the use of computers (Spiller, 2002: 84).

Price's interest in this notion of calculated uncertainty arguably developed most fully and systematically many years earlier, during his first major and best known project: the 'Fun Palace' (1961–1974). This was a scheme

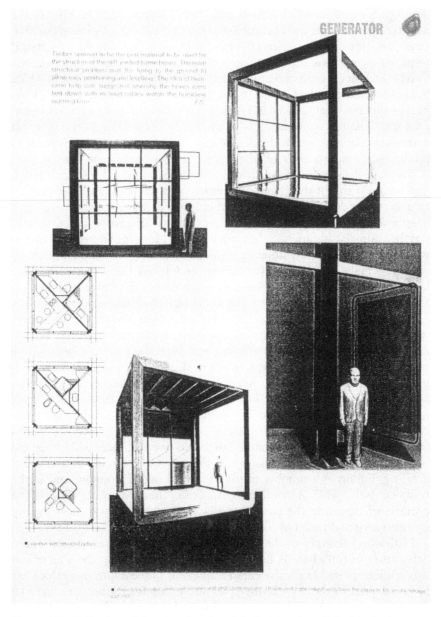

Figure 4.5   Collection of photographs of models showing the cellular blocks that were to form a core element in Price's proposed but unimplemented 'Generator' project.
Source: Cedric Price, architect, 1934–2003. Plans and perspectives of typical cubic units for Generator, 1976–1979. photomechanical reproduction, red ink and blue crayon. sheet: 41,6 x 29,5 cm (16 3/8 x 11 5/8 in.) DR1995:0280:621
© Cedric Price fonds Collection Centre Canadien d'Architecture/Canadian Centre for Architecture, Montréal. Reproduced with permission.

designed for and in collaboration with Joan Littlewood, a founder of the Theatre Workshop at the Theatre Royal in London (and a legion of collaborators and advisers). Littlewood's vision for the Project was for a theatrical venue where performance could flourish unconstrained by built form.

Although never built, it was intended as a deliberate experiment in which 'non-authoritarian ideas of use were to be tested' (Allford, 1984: 7). As Mary Lou Lobsinger (2000a: 24) puts it, it was to be a 'temporary, multiprogrammed [and reprogrammable] twenty-four-hour entertainment center that marries communications technologies and standard building components to produce a machine capable of adapting to the users' needs and desires'.

The Fun Palace was to be divided into various areas. These included: a 'fun arcade' ('full of the games and tests that psychologists and electronics engineers now devise for the service of industry or war'); a music area (with 'instruments available, free instruction, recordings for anyone, classical, folk, jazz and pop disc libraries . . . '); a 'plastic area' ('for uninhibited dabbling in wood, metal, paint, clay, stone or textiles . . . '); and a 'science playground' (which by night 'will become [a] *kaffeeklatsch* where the Socrates, the Abelards, the Mermaid poets, the wandering scholars of the future, the mystics, the sceptics and the sophists can dispute till dawn') (Littlewood, 1964: 432).

An integral component of the Fun Palace project and its various spaces was the inclusion of the latest in communications technologies, including reading and calculating machines, and televisions and computers. Key among these gadgets and a centrepiece of the scheme overall was the proposed 'Pillar of Information'. According to Stanley Mathews (2006: 45), this was to be 'an electronic kiosk that could search, display, and track information', and was 'among the earliest proposals for public access to computers in order to store information of all sorts'. As Littlewood explains in a 1964 article, 'at various points, sheltered or open, there will be screens on which closed circuit television will show, without editing or art, whatever is going on at a number of places in and out of London, and in the complex itself' (432). According to one critic, these technologies 'held the promise of thrusting the participant[-occupant] beyond mundane reality and into a virtual realm of communication' (Lobsinger, 2000b: 128).

Littlewood thought of the project as a 'university of the streets' and a 'laboratory of fun'. But, as Landau (1984: 11) points out, 'the idea of *fun* was not interpreted as passive entertainment as in the *amuse-me* ethic later to be adopted in the Walt Disney pleasure grounds'. Rather, 'it would be *fun* if the visitor could be stimulated or informed, could react or interact, but if none of these suited, had the freedom to withdraw' (Landau, 1984: 11). One contributor to the project suggested that the Fun Palace was less about 'fun' than it was about 'seeking the unfamiliar, and ultimately transcending it' (quoted in Mathews, 2006: 43). Littlewood (1964: 432) stressed that, 'the essence of the place will be its informality: nothing is obligatory, anything goes'. In short, Littlewood and Price are advocating social experimentation via technological engagement.

Price translated Littlewood's emphasis on the internal programmatic flexibility by designing the Fun Palace as an unenclosed steel frame structure with reconfigurable internal spaces and equipment (see Figure 4.6). Rearrangement was to be achieved by a permanent travelling gantry crane spanning the entire structure (Price, 1964: 433; 1984: 56–61). Freedom of interaction and withdrawal was also incorporated into the design through the structure's lack of enclosure and the omission of conventional doorways ('having no doorways enables one to choose one's own route and degrees of involvement with the activities') (Price, 1964: 433). What is more, it has also been suggested that 'there was to be no administrative hierarchy to dictate the program, form, or use of the spaces' (Mathews, 2006: 43).

Thus, in Littlewood's and Price's terms, the Fun Palace is a sort of huge 'anti-building'. Or, as one critic puts it, it is a 'vast mechanism that allows arrays of different kinds of space to be suspended in any position and continuously adjusted, moved or removed according to the changing needs of up to 55,000 simultaneous visitors (Wigley, 2004: 16).

It is important to note that the project was not just a fantastic proposal, the product of two people's fertile and unrestrained imaginations. Rather, it was developed with meticulous research and development between 1961

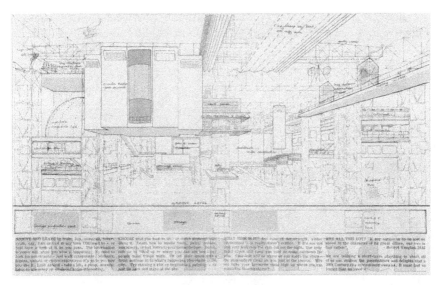

Figure 4.6 Internal elevation drawing of the Fun Palace. This image shows the various gangways, scaffold frame, and open design that gave the Fun Palace its 'unfinished' look.

Source: Joan Littlewood, author, 1914–2002; Cedric Price, author, 1934–2003; Cedric Price, architect, 1934–2003. Fun Palace: Promotional brochure, 1964. black and red ink reprographic copy on wove paper. sheet (unfolded): 36,2 x 59,8 cm. DR1995:0188:001:016

© Cedric Price fonds Collection Centre Canadien d'Architecture/Canadian Centre for Architecture, Montréal. Reproduced with permission.

and 1972, and was only shelved in 1974 after efforts to secure both a site and adequate financing failed in the face of various forms of opposition and 'lack of political imagination' (Shubert, 2004: 17).

It is worth remembering that Price was not the only one developing ideas of a technologised, ludic space for exploring alternative social forms at this time. For example, contemporaneous with the Fun Palace is the equally ambitious 'New Babylon' project by one-time Situationist International member and artist, Constant Nieuwenhuys, or Constant as he was generally known. Initiated in 1958 and stopped in the same year as Fun Palace (1974), New Babylon was Constant's long-term attempt to give form to the (Situationist) concept of 'unitary urbanism', according to which 'the urban environment [is understood] as the terrain of a game in which one participates' (quoted in Sadler, 1998: 120). Addressing similar ideas to the Fun Palace—especially play and flexibility—Constant presented New Babylon through an ever-expanding series of models (see Figure 4.7), knowledge of which was widely disseminated via exhibition and through publication. Indeed, it has been suggested that, 'any apparent affinity between the Fun Palace concepts of creative leisure and the creative strategies of the Situationist International would hardly be accidental, since both grew from common ideological and artistic roots' (Mathews, 2006: 41).

There are, however, several points of difference between Constant's New Babylon and Price's Fun Palace. A key point of difference rests in Price's

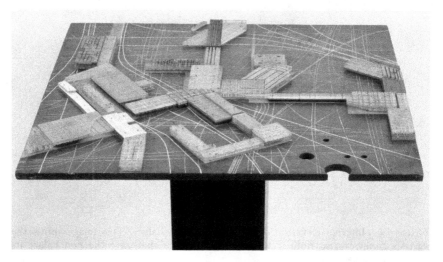

Figure 4.7   Photograph showing one of the many models developed by Constant as part of his long-running New Babylon project. This model shows the so-called 'Groep sectoren (Group of Sectors)'.
Source: Constant, Groep sectoren, 1959
© Collection Gemeentemuseum Den Haag. Reproduced with permission.

more explicit engagement with computing and communications technologies (although later models of New Babylon did incorporate electronics equipment) (Wigley, 1998: 66). Price's uses of these technologies were always closely aligned to and at the service of the social goals of the Fun Palace project. For instance, one proposal was to use a computer within the Fun Palace as a problem-solving technology to compile and sort data on user trends which, Mathews (2006: 44) explains, would 'set the parameters for the modification of spaces and activities within the Fun Palace'. Teletechnologies, then, were considered crucial to encouraging chance encounters and chance 'community'—the aforementioned 'eruption or explosion of unimagined sociality through pleasure'. Furthermore, they were crucial to the production of the social and the individual in Price's project, both physically and virtually. An even more apparent point of difference between Price's Fun Palace and Constant's New Babylon, and what connects Price's project to the earlier texts discussed in this chapter, is Price's guiding interest in systems theory and cybernetics. Indeed, a Cybernetics Committee was established and headed by Gordon Pask, the noted English cybernetic theorist and pioneer (Lobsinger, 2000b: 130).

Price's incorporation of these theories, and what sets this usage apart strikingly from their wider architectural application, is the marked difference between the research towards the Fun Palace and its intended patterns of occupation. On the one hand, guiding the project was a thorough, 'scientific' program of detailed research. On the other hand, the aim was for a ludic space, a space that encourages fleeting and unstructured processes of engagement. In other words, Price's approach bears comparison with the contemporary investigations of the Oulipo (the French-based 'workshop of potential literature') and its interest in rule-governed processes of production that liberate rather than constrain creative engagement. Mathews (2006: 41) argues that this is not at all surprising insofar as information technologies, cybernetics, and game theory 'are, in essence, means of modeling and systematizing chance and indeterminacy' (41).

Mary Lou Lobsinger develops a far more pessimistic reading of this union. She argues that the work of the Cybernetics Committee both expanded and problematised the goals of the Fun Palace project (Lobsinger, 2000b: 132). For Lobsinger (2000b: 132–133), the rather uneasy melding of a 'cybernetic learning machine' enabling 'self-regulation to achieve group consensus', and 'light-hearted pleasure seeking' takes the Fun Palace project into 'far stranger territory' than Littlewood had ever intended. Indeed, Lobsinger suggests that the result of this double influence is a somewhat 'nefarious kind of [social] control—invisible, apparently freeing and constraining at the same time' (2000b: 133).

These criticisms notwithstanding, it is still valuable to acknowledge the overall significance of the Fun Palace project and Price's work in general.

Significance of the Fun Palace

In architectural terms, the impact of the Fun Palace project has been profound. It simultaneously mounted a provocative challenge to monumental architecture, and rejected the unspoken conventions concerning 'respectable' architectural form and external aesthetic resolution of this form. But more than this, as one commentator remarks, the 'Fun Palace explored issues, crossed boundaries, and attempted to address social and political [concerns] that go far beyond the typical boundaries of architecture' (Shubert, 2004: 17). Remarking on the lasting significance of the theoretical coupling of architecture and cybernetics and systems theory in this project, Lobsinger writes:

> The cross-disciplinary process—tacitly based, as was the Fun Palace itself, on ideas borrowed from systems-design theory, especially that of self-organizing systems—may be its most significant contribution to recent architectural history and theory. (Lobsinger, 2000b: 133; see also Mathews, 2006: 41)

The project is also significant for its social ambitions and implications. Lobsinger argues that the 'production of the social and the individual—both physical[ly] and virtually—in real-time is the theoretical crux of the Fun Palace' (2000b: 128). This is borne out in one early, unpublished document by Price, where the stated objectives of the Fun Palace are 'to arrange as many forms of fun as possible in one spot [. . .] in the hopes of an eruption or explosion of unimagined sociality through pleasure' (quoted in Lobsinger, 2000b: 128).

Given these social ambitions, it is tempting to conceive of the Fun Palace as an architectural precursor to certain forms of on-line social space, particularly given Littlewood's emphasis on the Fun Palace as a media space; a space for information transfer; a space for education and pleasure; a social space; and a space where one is given the freedom to engage or withdraw. Such a comparison, however, is at once an overstatement and an oversimplification of both 'spaces'.

More useful is to consider what Littlewood and Price hoped to achieve in and through the creation of such technologically equipped spaces of and for experimentation. Lobsinger (2000a: 24) comments that, 'At the most literal level, activities such as the maneuvering of building components or group determination of program involve a basic form of social interaction'. But, additionally:

> It was also imagined that the Fun Palace would be equipped with the latest in communications technology: teaching machines, televisions and computers. These gadgets promised to thrust the participant-occupant beyond mundane reality into a virtual realm of communication. (Lobsinger, 2000a: 24)

In short, the Fun Palace was conceived as a 'giant learning machine with the capacity to enable humans to adapt physically and mentally to the intangible experiences and accelerated pace of technological culture' (24).

In the Fun Palace, questions of place and the changing exigencies of occupation and use are not only explored in tandem with questions of technology and its use: the two sets of questions are considered inseparable.

This is a key point of difference between Price's work and some of the other experiments in architectural mobility canvassed in Cook's *Experimental Architecture*. For example, while there is an emphasis on the ephemeral and the mobile in Price's work, 'place' is not seen as antithetical to technology (and vice versa). As noted earlier, 'The technological mandate [of the Fun Palace] moved beyond the realm of mechanical mobility', through the use of the gantry crane, and 'into the more ephemeral mobility offered by new information media and mass communications.' (Lobsinger, 2000b: 122) This has significance for how place was to be experienced in the Fun Palace. The integral role that telecommunications were to play in the scheme suggests a conception of place that is bounded but open and which is subject to and shaped by (among other influences) the flows of telecommunications and information.

While Price cannot be said to have resolved the 'problem of community' discussed in Chapter 2, his approach provides some rich material for contemplation and consideration, especially in relation to his (and his collaborators') conscious engagement with the interconnections and interactions of technology, place, and the social. Notwithstanding charges of social control, the Fun Palace project opens up some interesting points for comparison with the work of Herman Schmalenbach and Jean-Luc Nancy, along with some writing on CMC and virtual community, on unstable and fleeting forms of sociation. In social terms, a key contribution of Price's work—and of experimental architecture more generally—is its negotiation of the complex relationship of architectural design, technology, place, and 'community' through the creation of 'temporary flexible architectures for temporary flexible politics' (Galloway, 2006). Price's architecture, particularly the Fun Palace project, makes an important contribution to the history of telecommunications and human-computer interaction, and to performing urban spatial and cultural critique. Mathews (2006: 45) goes so far as to suggest that, 'The resultant web of information and free association to be produced by the proposed "Pillar of Information" [within the Fun Palace] not only anticipates the Internet by some three decades but also recalls the rhizomatic theories of knowledge developed in the 1970s by Gilles Deleuze and Félix Guattari'.

ARCHITECTURAL PROBLEM-SOLVING AND THE STRUGGLE WITH SOCIAL AMBITIONS

From the examinations of this chapter, a number of observations and points of significance have emerged concerning the use of computer technologies.

To begin with, the examples examined here show that the computer is taken up in complex ways that vary considerably from text to text. There is also an overall shift in the perception of these technologies, from initial scepticism or caution, to a more central role in architectural experimentation. For example, Chermayeff and Alexander view the computer as an aid for the architect, albeit one with quite restricted application. Negroponte, meanwhile, conceives of the computer as a powerful problem-solving technology that, one day, will become an intrinsic part of architectural fabric. This idea of the computer becoming intrinsic to architectural fabric is also canvassed, although in quite a different context, in Cook's survey of experimental architecture. Price, while not going quite as far as Negroponte, nevertheless also sees teletechnologies as a powerful tool for architectural and social experimentation.

In the cases where the computer is employed as a design aid, it is employed almost exclusively as a problem-solving technology and not as a form-generating device. What is interesting about the early non-representational approach to architectural computing is that it is the direct inverse of contemporary architectural computing. Here, computers (in tandem with computer-aided design software and coding such as virtual reality modelling language) are often employed primarily as a representational or form-generating technology.

Two further and key observations concern how community and place are understood in relation to these experiments in architectural computing.

In respect to community, all of the texts examined in this chapter purport to be guided by a 'humanist' agenda. Architectural computing is utilised in the resolution of social issues pertaining to housing and urban planning, with an aim to improve the human condition and the conditions of human habitation, yet there is very little detailed discussion of the social or of 'community'—at least in any detailed sense, or which extends beyond the provision for architectural choice. The most glaring example of this is Chermayeff and Alexander's *Community and Privacy*, which, despite its title, has no detailed discussion or even explicit mention of the notion of 'community' in the text. At very least, this serves as support for Adrian Forty's (2004: 117 & 103–117) point that (modern) architecture struggles to articulate its social ambitions. As detailed above, a key exception to this is Cedric Price, who seems to have little trouble articulating his social goals—particularly in respect to his PTb and Fun Palace projects.

Place, too, emerges as a notion that is of reduced importance in comparison to other considerations. Across all the earlier textual examples cited in this chapter, the dual emphasis on standardisation and individual choice leads to a retreat from questions of place (and, particularly, from Norberg-Schulz's (1980) influential consideration of *genius loci*, or 'spirit of place'). Again, the key exception is the work of Cedric Price, especially in his PTb proposal and Fun Palace project. In the latter case, place, as noted above, is understood as bounded but open and subject to, and shaped by, the flows of telecommunications and experimental social engagement.

The above highlights the point that, when it comes to architectural computing, a very different outcome is achieved when these teletechnologies form part of the *outcome* of the design process rather than just a *tool to aid* this process. Mark Wigley is more sceptical, and gives the following rather acerbic and pessimistic assessment of such work:

> Most of the idyllic images of experimental architecture were underpinned by electronics, even in the work of those who never addressed the subject. The absence of many traditional architectural elements in the images is made possible by the fantasized structural possibilities of the new electronic technology. (Wigley, 1998: 66)

While there is certainly some truth to this criticism, Wigley's assessment does not acknowledge the complex ways in which architectural structure, teletechnologies, and spatial use, all interconnect in this experimental architecture. It is the entwinement of these things, particularly in Price's work, which is considered by many to be still challenging today.

The final point concerns the influence on experimental architecture of systems theory and cybernetics. These influences are significant in that they show at least one wider theoretical and conceptual context in which these technologies are being deployed. As with the computer itself, here, too, the influences of systems theory and cybernetics varies—from scattered references to von Bertalanffy in *Community and Privacy*, to far more enthusiastic and deeply embedded engagement in the work of Price and Negroponte.

The influence of systems theory and cybernetics on experimental architecture is also noteworthy in that these same influences form a vital thread in the literature on, and critical engagement with, cyberculture and the internet (Tofts and McKeich, 1998; Keller, 2002). Moreover, the examinations of place, technology, mobility, and the social in this experimental architectural work (especially that by Price, and that described by Cook), run parallel with many of the key historical developments in personal computer manufacture and the emergence of computer-mediated communications capabilities. For these reasons, it is possible to view these early experiments in architectural computing as forming an important but hitherto neglected chapter in the broader history of cyberculture.

All of the texts examined above (and their wider influences and supporting evidence), drawn from different points in the early history of architectural computing, provide key insights into how these computing technologies were understood and valued for their design problem-solving possibilities. This material can also provide a solid basis from which to test and compare, in the next chapter, the ways in which computing technologies are imagined and discussed in a later phase of architectural writing on teletechnologies in the 1990s. This particular decade formed a key era of growth in the wider update and use of networking computing and the Internet (or cyberspace as it was then known). Architectural experimentation with teletechnologies during this time entered an even more ambitious phase, one in which these technologies were cast as the way forward, not just for architectural design, but for architecture as a whole field.

5 Fantasies of Transcendence and Transformative Imagination
Architectural Visions of Cyberspace

'Every great architect is—necessarily—a great poet. He must be a great original interpreter of his Time, his Day, his Age.'

—Frank Lloyd Wright (1970: 14)

'Formerly it appeared that experience left only one kind of space to logic, and logic showed this one kind to be impossible. Now, logic presents many kinds of space as possible apart from experience, and experience only partially decides between them. Thus, while our knowledge of what is has become less than it was formerly supposed to be, our knowledge of what may be is enormously increased. Instead of being shut in within narrow walls, of which every nook and cranny could be explored, we find ourselves in an open world of free possibilities, where much remains unknown because there is so much to know.'

—Bertrand Russell (1980: 86)

'What distinguishes human labor and the worst of architects from the best of bees is that architects erect a structure in the imagination before realizing it in material form.'

—Karl Marx (quoted in Harvey, 2000: 159)

Having detailed early, experimental engagements with architectural computing in the previous chapter, the focus now shifts to consider the enthusiastic embrace of cyberspace within the architectural literature of the early to mid-1990s. This literature on architecture and cyberspace consistently argues that networked computing technologies, imaging software, and conceptual understandings of cyberspace, create the conditions for a 'new' architecture of and for cyberspace.[1] In other words, computing technologies and associated computer-aided design (CAD) software not only extend the existing creative possibilities for architects and architecture, but also force a revised set of architectural understandings because they facilitate and enable a whole 'new' set of architectural possibilities.

This 'new' architecture of cyberspace is claimed by some to reveal the limits of 'conventional' architecture as it is presently known and understood, challenging many of the taken-for-granted assumptions about architecture and architectural construction. Thus, 'cyberspace architects' (if they may be so named) see themselves, to borrow the words of Frank Lloyd Wright (1970: 14), as great original interpreters of 'their Time, their Day, their Age'.

It is claimed that cyberspace architecture poses a radical threat to existing architectural design practice by overturning or dismantling many of its foundational elements—the semantic units that make up the very language of architecture. Such claims are borne out of a belief that a move from design and construction in physical space, to design and construction in the seemingly abstract realm of telecommunications networks and networked computing, enables a complete reconceptualisation of existing spatial rules and (thus, by extension) revolution for design grammar.

This chapter examines architectural writing on cyberspace to discover how cyberspace has been framed and understood in these texts—what themes emerge, what tropes and wider contextual frameworks and theories are used, and so on. It also maps what possibilities and opportunities cyberspace is seen to offer architecture and computer-aided design. Within this, a further aim—one that is core to this book—is to consider the position that the notions of 'place' and 'community' hold in these texts.

The approach involves a close reading of a select number of key texts. These include (but are not restricted to) the edited book *Cyberspace: First Steps* (Benedikt, 1992c), and two key themed issues of the journal *Architectural Design* on the topic of 'Architects in Cyberspace' (1996, 1998). The first of these, *Cyberspace: First Steps* represents an early collection of essays on cyberspace, and has come to be regarded as a seminal text in scholarship on cyberculture and cyberspace. Within this collection, the two explicitly architecturally focused contributions to this collection, by Michael Benedikt (the book's editor) and Marcos Novak, have also become benchmark essays for architectural writing on cyberspace. The second set of texts are drawn from *Architectural Design* journal, which constitutes a key forum for the publication of new and leading ideas in architectural theory and design, especially as they are aligned with new technologies and developments in science.

EXAMINING CYBERSPACE ARCHITECTURE'S KEY TENETS

The clearest means by which to explicate the claims that are made for 'cyberspace architecture' is to address in turn each individual architectural 'element' that this 'new' architecture is said to challenge. These 'elements', as I have formulated them, are: architectural representation; building materials; scale and proportion; spatiality and space use; time;

and architectural fixity. In considering these tenets, there is no assumed hierarchy for how these 'elements' are ordered (for example, the fact that time appears after space and space use does not suggest a privileging of one over the others).

Architectural Representation

The first claim made in the name of cyberspace architecture is that designing in cyberspace requires new forms and tools of architectural representation. It is argued strongly within much architectural writing on cyberspace that Leone Battista Alberti's representational tools, which have been crucial to the development of Western architecture, are deemed irrelevant and that the plan is dead as a mode of representation (Novak, 1995: 42). As William J. Mitchell (1999b: xii) puts it, the architectural embrace of networked computing leads to an 'opening up of a new design frontier', and, as a result, 'we will have to invent whole new languages for design in cyberspace'. The reasons for this are twofold. The first is largely a matter of pragmatics, in the sense that design and manipulation within the abstract three-dimensional world of computer graphics (especially computer-aided design programs and virtual reality modelling language) means the two-dimensional plan and traditional plan view are outdated. Further, cyberspace architects claim that the architect's ability to 'morph, mutate, and hybridise' (Spiller, 1998a: 51) three-dimensional representational images in the 'cyberspace' of computer software renders obsolete Louis Sullivan's now famous 'form follows function' dictum of architecture—a claim that will be returned to later.

The second, more generally polemical, view is that the impending 'death of the plan' is attributable to the failure of the discipline of architecture to keep up with the 'modern worldview'—a phrase largely synonymous in this context with complexity theory (McGuire, 1995; Novak, 1995). The static two-dimensional plan is taken as a reflection of Western architecture's preoccupation with form and structure, and its present failure to accommodate an architecture that might draw from complex systems and evolution, an architecture that 'works on the level of behaviour' (Ascott, 1998: 32). Novak offers the following summary of these claims:

> Plan, section, elevation, axonometric, perspective, traces of pigment held by the tooth of vellum, ruler and compass, were perhaps appropriate to the cycles and epicycles of Ptolemaic, Copernican, and Galilean universe, or even the ellipses of a Keplerian universe, but are completely impotent in arresting the trajectories of subatomic particles, or the shapes of the gravity waves of colliding black holes. Once this is observed, it can be readily seen that the plan is dead because its world view is obsolete. (Novak, 1995: 45)

Building Materials

The alleged migration from the physical to the virtual realm that is said to follow cyberspace also calls into question basic assumptions governing building materials, where the tools of construction are no longer seen to be bricks and mortar but networks and data. It is a shift that is captured in the popular aphorism 'bits replace bricks' (Pearce, 1995: 7; Spiller, 1998a; Mitchell, 1995, 1999a; Negroponte, 1995). Michael Benedikt theorises this shift as a movement away from 'materialised' architecture and towards 'dematerialised' architecture, while nevertheless recognising the contradiction that is inherent in proposing such a notion:

> Dematerialised? The reader may be surprised. What *is* architecture, after all, if not the creation of durable physical worlds that can orient generations of men, women, and children, that can locate them in their own history, protect them always from prying eyes, rain, wind, hail, and projectiles [. . .] durable worlds, and in them, permanent monuments to everything that should last or be remembered? (Benedikt, 1992b: 14)

The transient nature of cyberspace or 'dematerialised' architecture that is discernible in Benedikt's rhetorical question mounts a significant challenge to prevailing notions of architectural durability and monumentality—a point that will be developed later with regards to 'time'. For the present, however, it will suffice to take from the above passage an understanding that even to propose an 'abstract' or 'virtual' architecture arguably compromises popular conceptions (although not necessarily linguistic definitions) of the term architecture. This compromise is made all the more clear in the light of the claim that, in the future, building materials in cyberspace 'will expand to include speed, sampled image, auditory environments, non-linear hypertext worlds and cinematic spaces' (Romero, 1998: 46).

Rethinking architectural form and materials to incorporate speed and other factors, changes that are largely a result of networked telecommunications, lead some to suggest that existing planning domains require radical amendments, moving away from narrow, more traditional considerations of the built (physical) environment and towards planning schemes that account for the influence of global telecommunications networks and satellite systems (Ascott, 1995: 40).

Scale and Proportion

A further architectural 'element' said to be challenged by cyberspace is scale and proportion. Architecture of the physical realm (at least in the Western tradition) has long held the human form as its measure of 'desirable' scale and proportion. This system dates back at least to the time of the Ancient Greek philosopher Protagoras of Abdera, who famously declared that, 'man is the measure of all things' (Blackburn, 1996: 307; Jencks, 1995: 162). Architectural applications of this idea were proposed by Vitruvius in his *The Ten Books of Architecture* (*De architectura libri decem*, written before 27 BC), and further developed during the Renaissance (Capon, 1999a: 28–34). The so-called 'Vitruvian ideal' is perhaps best remembered in the rendering developed by Leonardo da Vinci in his image of the 'ideal man': a familiar sketch of a male body standing upright, inscribed in the space of a circle within a square, with the geometrical and metaphorical centres of all three figures located at the body's navel (Rasmussen, 1999: 115; Boyer, 1996: 75).

The desirability of the proportion system within architecture, as represented in 'Vitruvian man', is also based on the findings of the Pythagoreans and their arithmetical interpretations of nature (Blackburn, 1996: 311–312). These in turn have fed the development of the notion of a 'golden section' rule (see Rasmussen, 1999: 106 & 104–126). In Renaissance times, Leone Battista Alberti and Andrea Palladio both favoured the application of simple ratios derived from the above formulae (and musical harmonics), proposing floorplans according to patterns of 2:3, 4:4, 4:5, and 4:6 (Rasmussen, 1999: 110–111; Ackerman, 1990: 151). Meanwhile, in the twentieth century, Le Corbusier developed further permutations on the notion of the Vitruvian ideal through his principle of proportion called 'Le Modulor' (see Le Corbusier, 1956, 1958; Rasmussen, 1999: 106).

The point here is that architecture, throughout European history, has used the play of mathematical ratios in the formulation of 'correct' or 'harmonious' arrangement, and while these ratios may be informed in part by musical and mathematical theory, they are equally bound to the human form itself. In very broad terms, then, cyberspace architects are reacting to the premise that 'architecture up to contemporary times has been formed as a mimetic, allegorical, or metaphorical projection of the ideal proportions of man' (Boyer, 1996: 76). In stark contrast to this, cyberspace is said to mount a significant challenge to existing systems of architectural anthropometry (Spiller, 1995, 1998a, 1998b, Novak, 1995). One claim is that emergent spatial environments and architectures will ultimately explode the classical notion of the Vitruvian ideal of the body altogether: 'Architecture as we know it is to a large extent influenced by the scale of our bodies. In the future this scale will not remain consistent' (Spiller, 1998a: 65). More generally, claims for the 'explosion' of the Vitruvian ideal in the face of cyberspace are also connected to a host of broader debates concerning the reconfiguration of the body in cyberspace.

Architectural debates concerning the relationship of human to built form are by no means restricted to writing on cyberspace. A wealth of literature in the field of wider architectural theory addresses this very issue, albeit quite differently to writing on cyberspace (in particular, see

Ostwald and Moore, 1998; Vidler, 1990, 1992). Contemporary architectural discourse and practice (particularly throughout the 1980s) has been described as being characterised by acts of extreme violence perpetrated against the human body (Ostwald and Moore, 1998). Cited as evidence of this are the architectural projects of Bernard Tschumi, Coop Himmelb(l)au and Daniel Libeskind, projects that purportedly depict a 'morcellated' body (Boyer, 1996: 76–77), a body that is dismembered and ruptured. Ostwald and Moore—whose writing closely follows the ideas of architectural theorist Anthony Vidler—term this process of architectural disembodiment the 'violent mimetic tradition', suggesting its practitioners seek as a desired outcome, 'the destruction of a particular type of architectural culture; the Vitruvian [mimetic projection] system of anthropomorphic proportions' (Ostwald and Moore, 1998: 135; see also, Sampson, 1996; Negus, 1998).

For Ostwald and Moore, architecture's 'morcellated body' is attributable to major paradigm shifts in contemporary science:

> The body/event relationship in science has been steadily changing since Copernicus. Since that time the body has stopped being the central frame of reference and has become a component within the event [. . .]. (Ostwald and Moore, 1998: 43)

Illustrative of this shift is the gradual transformations affecting standard measures, from the pre-metric (anthropometrically determined) 'foot', through its metric replacement, to more recent changes governing the accurate determination of the metre. Charles Jencks retells the second of the two transformations, detailing the replacement of the metre, as established by the 1875 Treaty of the Meter (which fixed the inch at 2.54 centimetres exactly, as set by a platinum and iridium prototype), by a caesium clock, which has the basic measure of a second as the duration of 9,192,631,770 cycles of atomic radiation (so that a metre is fixed as the distance that light travels through a vacuum, which is $1/299,792,458$ of a second) (Jencks, 1995: 163). These developments are further evidence of a shift away from the human body as considered as the measure of all things in architecture—a departure that is further summed up by physicist Hans Christian von Baeyer's statement that, 'in the twenty-first century the atom will replace man as the measure of all things' (quoted in Jencks, 1995: 163). It is a position embraced by cyberspace architects.

Spatiality and Space Use

To question notions of proportion in turn raises questions of how we conceive of space. At the basis of such questioning lies the popular view that in cyberspace, 'architectural and spatial rules are broken, torn and folded to compress space' (Ostwald, 1996: 87; Benedikt, 1992a: 125ff). Again, as has been the case in the preceding discussion, these claims for spatial 'violations' draw largely from developments in contemporary science that espouse notions of complexity (especially post-Euclidean geometry and relativity), with its implication that 'whatever continuity we perceive in the world is a constructed illusion' (Novak, 1995: 44–45).

The immediate challenge this poses to conventional architecture is said to be at least twofold. First, the dialectic of solid and void, a foundational tenet of architectural design, is potentially transformed as a result of the 'unique' representational nature of cyberspace architecture. Advocates of cyberspace architecture eschew strict binary distinctions in favour of an 'architectural poetics' that embraces notions of fluidity, indeterminacy, and flux (Novak, 1995). Thus, the concepts of solid and void are replaced by the notion of an 'information field', in which 'the distinction between "what is" and "what is not" is one of degree, and there can be many sampling points between the two' (Novak, 1995: 45).

Secondly, the distinction that separates 'pure' space from 'habitable' space is also questioned. To couch this in architectural terms, the spatial (re)configurations wrought by cyberspace allegedly collapses one of—if not *the*—most enduring (if misunderstood) ideas of Western architectural design theory: the dictum 'form follows function', which is generally taken to mean that utility and beauty should merge (see Forty, 2004: 149–172 & 174–195). In its place, Novak (1995: 43 & 47) advocates the adapted aphorism 'form follows fiction', by which he and other like-minded critics claim that the 'form' of cyberspace may ultimately be closer to the spatial worlds depicted in the fictional work of writers such as William Gibson, Neil Stephenson, Greg Egan, Greg Bear, Jeff Noon, and others (Spiller, 1998a; Armstrong, 1999; Damrau, 1999; Novak, 1995), along with avant-garde artists, writers, musicians, and earlier architects interested in the 'fantastic' (Novak, 1992).

For advocates of cyberspace, architectural design that draws its inspiration from alternative design paradigms, such as science fiction, thus presents the possibility of an 'opening-up of a series of new spatial frontiers' and bodily possibilities (Spiller, 1998a: 7). However, privileging space over human occupation of it—which is arguably what the above proposals amount to—might spell potential danger for the place of the body in cyberspace. This is a development resisted strongly by some critics, who call on architects to 'reject the framing of cyberspace as an escape from embodiment and materiality' (Franck, 1995, 1998: 19; cf. Hayles, 1996: 3). A more moderate stance, one that is positioned somewhere between the two, is the proposal that, at very least, 'architects and urban designers of the digital era must begin by retheorizing the body in space' (Mitchell, 1995: 28).

Having said this, there is also a suggestion that the above debate is being waged on false ground. This is the view of the American architect and theorist Lebbeus Woods who points out that throughout its history, and contrary to popular belief (and teaching), architecture has a fundamental concern with the 'abstract qualities of space' and not with 'human presences within it': 'All designed space is, in fact, pure abstraction, truer to a mathematical system than to any human "function"' (Woods, 1996: 280; cf. Forty, 2004: 256–275).

This observation prompts Woods to call for a conceptual shake-up within architectural thinking, discourse, and pedagogy. 'The time has come,' he writes, 'for architects to accept the essential emptiness of space, its voided meanings, its indeterminacy and uncertainty' (Woods, 1996: 285). He advocates this so as not to foreclose in advance any unlikely, unexplored or hitherto repressed spaces and spatial possibilities—principles that are synthesised in his interconnected notions of 'freespaces' and 'heterarchy' (see Woods, 1996). The pursuit of these principles lies at the heart of Woods's own architectural practice and his interest in cyberspace. Woods's observations on 'spatial emptiness' adds an important qualification to the present discussion. On the one hand, if Woods is to be believed, the collapse of the form/function divide within cyberspace might be read not so much as posing a radical threat to conventional architecture but as a return to its very origins. If so, it is a 'return' that is informed by a science that is interested in theories of chaos and complexity. On the other hand, Woods's position can also be read as a further expulsion—or 'voiding'—of the body in cyberspace, and thus further cause for concern for critics who resist the framing of cyberspace as a 'disembodied' space.

Time

A further broader philosophical tenet concerning cyberspace architecture pertains to time, where it is argued that the temporal dimension or qualities of this architecture constitutes a 'radical' break from existing architectural practice.

Marcos Novak traces this aspect of cyberspace back to the development of post-Euclidean geometry, isolating as a key moment in its history Hermann Minkowski's proposal of a 'fourth dimension, uniting space and time into space-time'—a proposal subsequently developed by Albert Einstein in his formulation of the theory of relativity and by more recent scientists, such as Werner Karl Heisenberg and Kurt Gödel (Novak, 1996). For Novak (1996), the central dilemma to emerge from this work for architecture is that, 'the period between the development of non-Euclidean geometries and the present has been a period of growing rupture between how we know the world and how we express that knowledge architectonically'. In pure design terms, this rupture is seen as acute. Novak suggests that architects are familiar with the problem of adding a third dimension to a two-dimensional plan. However, 'now that the plan is dead', the question is, 'how does one add a fourth dimension to the third?' (Novak, 1996) For Novak at least:

> Such a question challenges all aspects of our discipline. There are no theoretical, practical, or pedagogical models for such explorations—which

is exactly the reason why such explorations are interesting and important. (Novak, 1996)

Early experiments in adding a 'fourth'—or more accurately in this case, a 'temporal'—dimension to architecture were and continue to be explored by the British architect, John Frazer, a pioneer of cyberspace architecture. Frazer (1995) has spent the last four decades developing a 'new model of architecture' that is both inspired by and utilises the principles of evolution, such as 'natural selection' (20), 'self-organisation' (13), and other 'generative tools' (24) that affect change over time. This 'new model' Frazer labels 'evolutionary architecture', placing it within an 'emerging field of architectural genetics' (10). Perhaps the best known outworking of this theory of architectural genetics in practice was Frazer's collaboration with the architect Cedric Price on the (1978) 'Generator' project, discussed in the previous chapter (see Frazer, 2003).

Despite such experiments, for Novak (1996) at least it is really only in the virtual realm that 'suitable architectonic expression' can be achieved that will mend the 'rupture' he argues is afflicting architecture. This is based on the assertion that cyberspace architecture, when conceived of as a 'field of information', is 'designed as much in time as in space, changing interactively as a function of duration' (Novak, 1995: 45).

Conceiving of cyberspace architecture in these terms, it is proposed, suggests at least two consequences. First, Novak argues that the separation of the 'art of space' (architecture) from the 'art of time' (music) is no longer tenable. Their intersection and convergence led Novak to coin the neologism: 'archimusic'—an architecture purely of the imagination (1995: 44). Of course, Novak is not the first to make a comparison between architecture and music; there are, for instance, the musically inspired ratio systems of Alberti and Palladio, mentioned earlier. There is also the German Romantic philosopher Friedrich von Schelling's description of architecture as 'frozen music' (Von Schelling, 1989: 166; Martin, 1994).[2] However, there is a significant difference between the two statements by von Schelling and Novak. According to the latter, the distributional, durational nature of cyberspace means that architecture as music need no longer be frozen but 'liquid': a 'symphony in space, but a symphony that never repeats and continues to develop' (Novak, 1992: 251). The second consequence flows from the first: 'duration' means that conventional notions of architectural monumentality require revision. The idea here is that architecture ceases to be 'monumental' because it is transformed continually.

Fixity of Architecture

One alleged implication of the above is the 'collapse' of a further foundational tenet of architecture, and the very thing that makes buildings so much a part of the physical landscape. This is the very fixedness of architecture, the creation of 'durable physical worlds' and 'permanent monuments' (Benedikt, 1992b: 14). This is not to say that architectural fixity has hitherto remained unchallenged. Not only has architectural 'fixity' often been negotiated in tension with questions of architectural 'mobility', but these tensions regularly resurface, prompting renewed consideration. For example, these tensions were issues of pressing concern for experimental architects of the 1960s and 1970s, as was highlighted in the previous chapter in relation to the work of Peter Cook and Cedric Price (see also, Damrau, 1999; Dessauce, 1999).

In relation to cyberspace architecture, however, there is said to be a key difference that separates it from these earlier experiments in architectural mobility: the claim that the spatio-temporal nature of cyberspace and computer connectivity enables place *and* space to be distributed (Novak, 1995: 43; Mitchell, 2000). As one writer rather extravagantly puts it, cyberspace has enabled 'the first global nomadic mnemonic architecture' (Berry, 1995). It is these distributional, fluid qualities of cyberspace that lead Novak (1992, 1995) to preference the terms 'liquid architecture' and 'transmissible architecture' over the apparently more prosaic and descriptive 'cyberspace architecture'. Importantly, though, and apropos of Frazer's notion of 'architectural genetics' discussed earlier, Novak adds that 'what must be transmitted is not the direct object itself but [. . .] the *genetic code* for the regeneration of the object at each new site' (1995: 45—emphasis added). Similar sentiments are echoed in the introduction of the notion of 'vacillating objects', coined so as to distinguish cyberspace from a physical landscape marked by fixed or permanent objects (Spiller, 1998b). Implicit in this understanding is the idea that, because of cyberspace, the practice of architecture is increasingly removed from its more traditional preoccupation with the design and construction of static buildings. In its place are a series of preoccupations that would seem to share more in common with *rheology*: a medical term denoting the science dealing with the flow and deformation of matter, and its child, nanotechnology (Spiller, 1995; Meyer, et al., 1999).

CYBERSPACE ARCHITECTURE: A CRITICAL REFRAMING

Given the claims outlined above, how might we make sense of, and begin to critique, cyberspace architectural theory and its claims as a coherent whole, as a working system? The following discussion explores some of the motivations behind these visions of, and arguments for, cyberspace architecture. It does this in order to begin to understand architectural interest in this medium. Why has there been such a willingness to celebrate cyberspace as a 'liberatory' realm? More precisely, what fuels the desire to realise the possibilities this technology is seen to offer? Why is it so important that this 'space' constitutes a 'radical' challenge to present architectural understanding? And, lastly, how accurate are these claims to radicalness?

In order to respond to these questions, the following discussion considers the field as a coherent whole. After all, what is a 'theory' but a *complex* or *system* of ideas? The issues this raises for how we might then more fully understand cyberspace, architecture, 'place', and 'community', will then be examined.

One way to give form to emerging theoretical developments is to grant them a name, as we have learnt through Novak's coining of the terms 'liquid' architecture and 'transmissible' architecture. Approaching naming as a way of prying open or giving form to emergent ideas, I here propose the term 'neuromantic architecture' as a useful critical term for 'making sense' of the overall field of cyberspace architecture.

A 'New' Romanticism: Cyberspace Architecture as 'Neuromantic' Architecture

It should be noted at the outset that the creation of any new term risks a number of conceptual pitfalls. A useful rule of thumb, it would seem, is to ask the following of any new term: is it necessary? Does it add anything to present discourse? In this instance, as there are a number of pre-existing alternatives—such as 'dematerialised' or 'cyberspace' architecture (Benedikt, 1992a, 1992b), 'liquid' architecture (Novak, 1992), 'transmissible' architecture (Novak, 1995), 'electrotecture' (Bell, 1996), 'digital' architecture (Memarzia, 1997), and 'datatecture' (Hennessey, 2000)—why not persist with them? There could be some merit in persisting with one (or more) of these terms—such as 'cyberspace architecture'—but perhaps placing it 'under erasure [*sous rature*]', as Derrida (1997a: 60) might advise, in order to designate its inherent shortcomings. Putting a word under erasure, rather than coining a new term, can be useful because, as Gayatri Spivak (1997: xv) reminds us, 'to make a new word is to run the risk of forgetting the problem or believing it solved'.[3]

Alternatively, if one were to persist with the coining of a new word, it might be equally meritorious to propose it but immediately place it, too, '*sous rature*'—as in ~~neuromantic architecture~~—in order to highlight its own predestinate inadequacies.

Like word creation and naming, the act of 'grouping together' is equally problematic. In this case, to 'encapsulate' or 'group together' a variety of writings on cyberspace architecture under one term or rubric is to risk presuming that there is a totalising concept of cyberspace architecture, that each instantiation belongs to a certain *genre* of architecture (illustrative of this 'risk' is the fact that Charles Jencks has never explicitly addressed cyberspace architecture, but his writing on architecture and science nonetheless share marked similarities with those critics that do). For in naming it, it would presume to have enclosed (delimited, closed off) various (and sometimes conflicting) conceptual, theoretical, and political differences. The basic point of conventional genre criticism, however, has in general been to group like with like, according to a set of *shared characteristics*, which does not necessarily mean the same thing as *negating difference*—a point argued convincingly by Rick Altman (1989) in relation to film genre criticism. Thus, if indeed we can talk about cyberspace or 'neuromantic' architecture as a 'genre', the point is not necessarily to delimit this theory or deny difference, nor is it to portray it as monolithic. Rather, the intention is to show that there *are* a number of marked similarities in the way cyberspace architecture is conceived of and written about that shed light on the whole field, while not necessarily denying possible differences within this field.

Therefore, in response to the suggestion to continue with pre-existing terminology, I persist with this alternative term, *neuromantic architecture*, for the precise reason that it is not meant as a replacement for the earlier terms (as Spivak warns against) but as a supplement to them, allowing for an additional way of making critical sense of cyberspace architectures. As for the issue of writing '*sous rature*', it is duly acknowledged that the notion of neuromantic architecture is provisional. However, rather than render this term in 'strikethrough' font here, the nature of its own inherent shortcomings will emerge through the ensuing arguments and later, concluding chapter discussion.

Neuromantic architecture is not intended as a prescriptive or proscriptive term. Rather, it is primarily descriptive, and carries with(in) it a two-fold resonance.

First, it acknowledges William Gibson's (1993a) novel *Neuromancer* (which was one part of what he called the matrix trilogy, and which was first published in 1984) as a foundational text for cyberculture in general and cyberspace architecture in particular. It also acknowledges the use of the term 'neuromanticism' by other earlier critics, albeit in slightly different contexts to discussions of cyberspace architectural theory (Glazer, 1989; Csicsery-Ronay, 1988; Voller, 1993). While not intentional, it is also possible to discern pathological connotations in this term, as 'neuroma' refers to any disease of the nervous system. In the context of Gibson's novel, this is quite fitting, given the neural damage that Case, the book's main protagonist, suffers from his past explorations in cyberspace.

Secondly, and crucially, the term is evocative of the way in which cyberspace architecture, in response to claims for its intrinsic 'radicalness', can be read alternatively as a reconceptualisation of certain notions within the literary tradition of Romanticism.

Such a comparison between neuromantic architecture and Romanticism emerges from the possible connections between cyberspace architectural theory and the literary Romantic tradition based on the available architectural literature on cyberspace. This connection responds to the title of Gibson's novel and its allusions to the Romantic literary tradition, as well as acknowledging the power of puns (Ulmer, 1988) and the value of coincidence (Derrida and Kipnis, 1987: 172) in the organisation of knowledge.

What follows, then, is a critical reframing of architectural writing on cyberspace which (re)reads these texts as a return to certain elements of, or themes in, the literary Romantic tradition. The specific focus here is in cyberspace architectural theory's interest in the sublime, and its investment in the notion of transcendence and the transformative powers of the imagination. The term 'neuromantic architecture' does not necessarily replace other terms; it carries the above connotations more self-consciously.

The argument that cyberspace architectural theory represents a return to elements of the Romantic tradition hinges on the extent to which *Neuromancer* can be considered a Romantic text. For example, the American cultural critic, Miriyam Glazer, drawing on the work of Bruce Sterling, reads Gibson's fiction as 'romantic', insofar as it rejects science fiction's traditional 'love affair with Big Science' in which, 'the "technolatry" of earlier science fiction [. . .] was also a modern re-enthronement of the Enlightenment Goddess of Reason' (Glazer, 1989: 156). In place of this, Glazer argues that Gibson favours a view of science as intimate and pervasive, whose 'palpable presences [are] influencing every aspect of life' (155). In so doing, Gibson is said to privilege the imagination in the sense that, in his novels, 'rebels [fighting] against the dominant order are obsessed with experiences that sweep them up beyond the confines of self and where a desire to confront the unknown becomes an urgent force' (Glazer, 1989: 156). But, for Glazer, this is where the similarities between traditional literary Romanticism and Gibson's novels end. A profound difference lies in the endpoint of these experiences, which Glazer delineates as the 'optimism of Romanticism' versus 'the resignation of Neuromanticism':

> The earlier Romantics saw the journeys—spiritual, psychological, imaginative—on which they persistently embarked as opening up new horizons for the living. In Gibson's world, this Romantic faith in the inner life and, with it, the human imagination, as wellsprings of positive human and social transformation, have all but disappeared [. . .].
> (Glazer, 1989: 158)

One problematic aspect of Glazer's reading of literary Romanticism is its apparent neatness, where the power of the imagination is seen as an overwhelmingly 'positive' force. This was not necessarily so, as Abrams (1968) points out. To cite just one example, the sense of hope that was such a strong element in Romantic writings sprung at least in part from the Revolutionary happenings of the 1790s, but later took on a more apocalyptic tone of disillusionment (Abrams, 1968: 53). Thus, it is somewhat disingenuous to separate Romanticism and 'Neuromanticism' according to such a strict (and neat) dichotomy. Nevertheless, Glazer elides such considerations, founding her bleak assessment of 'Neuromanticism' on Gibson's particular take on technology and transcendence.

Of course, the theme of transcendence is central to much Romantic writing, with 'the paradigmatic Romantic journey leading to and reflecting a liberating expansion of consciousness' (Glazer, 1989: 160). For Glazer, the difference between Keats, Wordsworth, and company and the more recent writings of Gibson is marked. Compared with the work of the earlier Romantics, Glazer (1989: 161) argues that Gibson's 'cyberspace is not visionary or imaginative territory'.[4] Rather, it is, to quote Gibson (1993a: 67) himself, the 'nonspace of the mind'. The outcome of this is that the union between the natural and the human worlds—which Glazer, quoting Frye (1968), suggests is 'one of the major functions of poetry itself'—is denied in Neuromanticism (Glazer, 1989: 162). Instead, 'in Gibson's fiction the non-human world of the old paradigm, the world of nature, is decaying and peripheral' (162).

It is difficult to deny the overall dystopic flavour of *Neuromancer*. Indeed, Gibson's *oeuvre* can be read as a sustained and devastating series of literary pathology reports detailing the impact of technology on both urban and human form. In this context, it is noteworthy that Gibson cites Mike Davis's (1991) *City of Quartz* as the inspiration behind his own (1993b) novel *Virtual Light*—Davis's book being a damning critique of the history of urban development and social control in Los Angeles.

However, a major paradox lies in the way Gibson's novels, especially *Neuromancer*, have been received within cyberculture and cyberpunk, and particularly within cyberspace architecture. Whereas the English Romantics rallied against mechanisation and technology (Frye, 1968; Lucy, 1997: 44), it is the very wellspring of creative inspiration for cyberspace architects. Technology (and science) provides for cyberspace architects, to borrow Abrams's words, an 'intoxicating sense that now everything is possible' (Abrams, 1968: 33), just as the French Revolution did initially for the English Romantics.

But how can this be? If Glazer's summary is credible, how can such a desolate picture depicted in *Neuromancer* also provide such impetus for creative inspiration?

A large part of the answer lies in the sublime nature of Gibson's definition of cyberspace:

> Cyberspace. A consensual hallucination experienced daily by billions of legitimate operators, in every nation, by children being taught mathematical concepts [. . .] A graphic representation of data abstracted from every computer in the human system. Unthinkable complexity. Lines of light ranged in the nonspace of the mind, clusters and constellations of data. Like city lights, receding [. . .].
> (Gibson, 1993a: 67)

Commenting on this passage, Scott McQuire (2002: 168) observes that 'both the fragmentary nature of the definition and Case's [the central protagonist's] dismissive response immediately align cyberspace with the sublime, a domain of "unthinkable complexity" which eludes the nets of representation'. Here McQuire touches on the fact that what architects and architectural theorists are so enamoured of in Gibson's conception of cyberspace is the spatial possibilities that it evokes and registers, and the idea of a domain of 'unthinkable complexity', a 'space' of possibilities, that eludes representation.

This evocation of 'unthinkable complexity' is also connected strongly with the notion of 'postmodern art'. In a discussion of postmodern literary theory, Niall Lucy defines 'postmodern "art"' as the moment where 'virtual (postmodern) communities can recognize art only in terms of what Kant called "an outrage of the imagination" [. . .] through an experience of the sublime' (Lucy, 2000: 9; see also, Crowther, 1995). Accordingly, '"art" [. . .] could only be what *lacks* beauty', what is 'atrocious' (Lucy, 2000: 9). Or as Geoffrey Scott puts it, the Romantic sublime, 'identifies beauty with strangeness' and the unfamiliar (quoted in Capon, 1999b: 302). Importantly, flowing from—or more precisely, lying behind—this understanding of 'atrocious beauty' is the issue of how one thinks about the term 'nature'. This issue will be addressed at length below, as it is crucial to how cyberspace architecture can be understood as a 'return' to certain key ideas within the Romantic tradition.

Nature and Neuromantic Architecture

The 'nature' of Romanticism can take many forms, ranging from images of the pastoral idyll to more complex, 'sublime' visions of beauty, grandeur, and horror. More bucolic approaches are adopted by those such as Glazer, though as argued above, cyberspace architects have, for the most part, been stimulated by and followed closely Gibson's engagement with the 'postmodern' sublime.

Nature, in both these broadly 'Romantic' manifestations, has long proven an inspiration to architects, from John Ruskin and his contemporaries, through to practitioners such as Frank Lloyd Wright. Indeed, Wright has been described as 'one of the champions of Romanticism in twentieth-century architecture' (Capon, 1999b: 303) in that he eschewed classicism in favour of an 'organic architecture', advocating an 'architecture of nature, for Nature' (Wright, 1970: 3; see also, Frazer, 1995: 10; Steadman, 1979).

For architects of the virtual realm, 'nature' continues to be an inspiration (Dollens, 2005; Estévez, et al., 2003). However, neuromantic architects have, to borrow the words of Charles Jencks (1995: 165), 'returned to the nature that the Romantics loved but find it is quite different'. There are two reasons for this.

First, as has been stressed throughout the preceding discussion, neuromantic architecture draws much of its inspiration from nature as seen through the lens of contemporary scientific theories of complexity and chaos, through which the world is viewed as messy and complex. Nevertheless there remains a sense of continuity with the Romantic conception of nature. For the Romantics, as Lucy points out (quoting the British essayist Isaiah Berlin), 'the tidy regularities of daily life are but a curtain to conceal the terrifying spectacle of true reality, which has no structure' (Lucy, 1997: 47). Thus, as Lucy summarises, for the Romantics, '*lack* of structure therefore was positive, and it was natural' (47). This goes some way towards explaining why it is that many neuromantic architects suggest that taken-for-granted, conventional architectural 'construction-conceptions' (McGuire, 1995) need to be rethought in the light of complexity theory—a point that also emerges regularly in writing on cyberspace and architecture (see, for example, Benedikt, 1992a, 1992b; Novak, 1992; McGuire, 1995; Spiller, 1998a).

Secondly, such a 'different' view of nature comes about at least in part as a result of seeing nature 'in terms of processes' (Hunt, 1998: 54), with architects drawing inspiration from 'the inner logic of its morphological processes' (Frazer, 1995: 10). This particular take on nature is borne out of an interest in the field of cybernetics, which, as Hunt explains, 'can be thought of as an attempt to understand machines through analogies to organisms', specifically '"self-regulated" (homeostatic)' machines which have 'autonomous control over [their] own behaviour' (Hunt, 1998: 54).

Neuromantic architectural applications of these ideas originate in, and repeatedly return to, the work in the late 1960s of the English cybernetic theorist Gordon Pask (1969). Pask (1969: 494) viewed cybernetics and architecture as enjoying an 'intimate relationship', sharing a 'common philosophy' of 'operational research'. His argument develops the idea that 'architects are first and foremost system designers who have been forced, over the last 100 years or so, to take an increasing interest in the organizational (i.e. non-tangible) properties of development, communication and control' (494). It is a concern that also links 'architectural cybernetics'— and, by extension, neuromantic architects—directly with architectural interest in the systems approach and General System Theory.

As with architectural interest in the systems approach and early computing, Pask's interest in cybernetics is resolutely humanist in the sense that he sees it as a means of overcoming problems associated with human interaction (494, note 1).

Needless to say, the 'electronic' possibilities of Pask's ideas were quick to be realised. While not necessarily true to his original expression (or humanist aims), the biological analogies that lie at the heart of architectural cybernetics found fertile ground in the emergent field of cyberspace or neuromantic architecture (see Hunt, 1998; Gage, 1998; Ascott, 1998; Frazer, 1995). The challenge that this 'biological' or 'process' model of (cyberspace) architecture poses to conventional architecture (canvassed in the earlier key tenets section) finds its clearest articulation in Roy Ascott's declaration that, 'the problem with Western architecture is that it is too much concerned with surfaces and structures and too little concerned with living systems' (Ascott, 1995: 39). Addressing this problem has been the impetus behind Frazer's 'evolutionary' and Novak's 'transmissible' models of architecture.

It has also given rise to a number of more elaborate applications of the 'biological analogy'. One proposal is to suggest that conceiving of neuro-mantic architectures ('technoetic structures', according to Ascott's termi-nology) as a living system is inevitable in a society characterised by global informational networks that exhibit similar biological traits of feedback, adaptation and change (Ascott, 1995, 1998). Ascott (1995) gives the name 'cyberception' to the dominant modes of perception and understanding of this societal matrix. As he describes it, 'cyberception involves a convergence of cognitive and perceptual processes in which the connectivity of telematic networks plays a formative role' (1995: 38). Despite the rhetoric, however, Ascott's theory of 'cyberception' is ostensibly a variation on McLuhan and Fiore's (1989: 40) familiar notion of the media as extension of the human central nervous system.

A further suggestion is that the recent trend for so-called 'smart' or 'intelligent' building is the logical outworking of this biologically driven approach to architecture, where building materials are seen as 'living', as part of a 'responsive system' that brings about 'greater environmental adaptivity' (Hunt, 1998: 54–55)—as can also be seen in the architectural experiments of Cedric Price discussed in the preceding chapter. Architec-ture of this sort is seen to be connected to human—or strictly speaking, *biological*—systems, where the 'actuators behave like muscles, sensors serve as nerves and memory, and the inclusion of communications and computational networks represent the brain and spinal column' (Hunt, 1998: 55). And, like Pask's theory of architectural cybernetics, the idea of this sort of 'soft, responsive architecture' (Frazer, 1995: 17) has found fertile ground among architects, attracting many variations on a theme with many different labels. There is 'recombinant architecture' (Mitchell, 1995, 1999; Nixon, 1996), 'cooperative buildings' (Kaplan, Fitzpatrick and Docherty, 2000), 'intelligent spaces' (Riewoldt, 1997), 'smart buildings' (Negroponte, 1995; Betsky, 1999: 168–173), and so on. Frazer's work on Cedric Price's 'Generator' project is seen as the forerunner to these ideas. Ironically, though, Frazer prefers to distance himself from them, lamenting the 'devaluing' of the term 'intelligent building' from this earlier project to what he sees as its contemporary meaning as little more than 'building with information technology' (Frazer, 1995: 41).

But all of this is only part of the picture. Adequate explanation of the central place of 'nature' in neuromantic architecture also hinges on how one circumscribes the term 'nature'. To see 'nature' as 'dead' in *Neu-romancer* (as Glazer and others do) is, to a certain extent, to conceive of nature as singular. This is not necessarily the case. For instance, the cultural critic McKenzie Wark argues persuasively that 'nature' can hold tertiary meaning (at least). Wark describes his 'layered' understanding of nature in the following terms:

> From the telegraph to telecommunications, a new geography has been overlayed on top of nature and second nature. [. . .] Second nature, which appears to us as the geography of cities and harbours and wool stores is progressively overlayed with a third nature of information flows, creating an information landscape which almost entirely covers the old territories. (Wark, 1994: 120; see also, Wark, 1990, 1993; Tofts and McKeich, 1998: 22–23)

Much of what Wark discusses here is discernible in the three applications of the 'biological analogy' to architecture mentioned directly above. But it is also significant that, in developing this theory of 'third nature', Wark links it explicitly with Gibson's *Neuromancer* in the sense that he suggests 'cyberspace in literature' and for architecture is virtually synonymous with 'third nature' (Wark, 1993: 17). The implication is that 'occupants' of 'third nature', like many of the characters in Gibson's fiction, 'inhabit' not 'the actual terrain in which we live and work and play, but rather the virtual space of media flows that come to pass through the eye and the ear and nestle in the unconscious' (quoted in Adams, 1996: 35).

This is not to say, however, that the much-claimed 'death' of 'first nature' is ignored by neuromantic architecture. On the contrary, the terminal condition of 'first nature', that Glazer talks of, haunts much architectural writing on cyberspace; it also lies at the root of some of the (admittedly less convincing) claims for digital technologies as a panacea for environmental destruction (see, for example, the final chapter of Mitchell, 1999a). What the notion of 'third nature' does offer is a way of understanding why it might be that neuromantic architects hold such store in Gibson's novel as a source for creative inspiration. In a word, the neuromantic imagination is fired by 'third nature' and its invocation of a 'tantalizing abstraction' of a '"space" that has no co-ordinates in actual space' (Tofts and McKeich, 1998: 15).

To synthesise the above material then, neuromantic architects draw their inspiration from a threefold conception of nature: a nature influenced by scientific theories of complexity, cybernetics, and the 'third nature' of global informational flows.

Imagination and Neuromantic Architecture

To understand nature in this way not only is to allow a reinstatement of the imagination into the neuromantic project but is to recognise its centrality in that project. This is due at least in part to a conception of cyberspace by neuromantic architects and other techno-boosters as a space of possibilities. These sentiments mirror those of the English analytic philosopher Bertrand Russell in his short book, *The Problems of Philosophy*. Writing on the limits of philosophical knowledge in the comprehension of spaces, Russell remarks, that 'we find ourselves in an open world of free possibilities, where much remains unknown because there is so much to know' (1980: 86). While written well before the age of cyberspace (*Problems* was first published in 1917), Russell's *compte rendu* (the full passage is included as an epigraph to this chapter) nevertheless reads like a kind of dictum for neuromantic architecture and cyberculture generally.

In order to realise these potentialities, and in order to tap into the cyberspatial 'unknown', there is a suggestion that, like McGuire's (1995) call for rethinking 'construction-conceptions' in the light of complexity theory, architects should re-evaluate the applicability of existing metaphors to cyberspace. This is because, 'the immediacy and relative ease of transplanting reference models and metaphors from our memories in physical reality can only provide the false comfort of association, which in the end only masks the indiscernible turbulence of technological change' (Romero, 1998: 50). Rather, it is preferable to avoid complacency by approaching the 'unknown' of cyberspace, as much as is possible, unfettered by existing associations and paradigms.

An alternative way to describe this process, to adopt a phrase employed in another context, is as a 'defamiliarising of the ordinary' (Tofts, 2000b: 14). This was (and is) an established strategy within the avant-garde arts, including Russian Formalism (O'Toole, 1994: 272, note 20). It was especially strong within Surrealism, where the imagination—in addition to other 'forces', such as the subconscious—was granted free reign as a sort of creative Golconda.

Neuromantic architects likewise place great stock in the imagination. But it is in the imagination as saddled to technology—employed as a master key of sorts for unlocking, in this case, the cyberspatial 'unknown'. The sentiment behind this, as captured by Robert Harbison (1992: 168), is that, 'better than ordinary works of architecture enshrine the metaphysical approach to technology, which sees it as a visionary medium for realizing the most far fetched dreams'. Thus, to the extent to which digital technology might be described as a 'visionary medium', neuromantic architecture is an attempt to realise the most far fetched dreams—realising what was hitherto unrealisable in architectural terms. In this sense, neuromantic architecture falls under the rubric of 'fantastic' architecture—a term developed by Ulrich Conrads and Hans Sperlich (1963) in their study of speculative architectural design proposals and unbuilt (and largely unbuildable) buildings—or what Harbison (1992) calls 'paper-based' buildings (see also, Damrau, 1999; Wrede, 1990; Cooper, 1996). Conrads and Sperlich (1963) describe fantastic architecture as a form of fantasy, an attempt to 'pay tribute to human imagination; specifically, to that creative fantasy which conveys its message [. . .] in architectonic shapes' (7).

Of course, it almost goes without saying that the creative impetus of the twin notions of 'fantasy' and 'imagination' stretch well beyond the confines of Conrads and Sperlich's book, covering all forms of creative output, whether literature, art, or architecture. On the last of these—architecture— Anthony Antoniades comments that, 'the concepts of *fantasy* and *imagination* are [. . .] paramount as ingredients in the creative process' (1992: 6—original emphasis). For instance, it was Alberti who suggested that 'we can in our Thought and Imagination contrive perfect Forms of Buildings entirely separate from Matter' (quoted in Novak, 1992: 244). In 1919, Walter Gropius advised architects to 'build in the imagination, unconcerned about technical difficulties' (quoted in Novak, 1992: 247). And, in 1968, the Viennese architecture studio of Coop Himmelb(l)au took their name because it referred to 'imaginative architecture'—'the idea of having architecture with fantasy, as buoyant and variable as clouds' (Werner, 2001; Jencks, 1995: 161). Indeed, it is Marx who suggests (albeit rather unflatteringly) that this imaginative capacity is a crucial aspect of what makes us human. 'What distinguishes human labor and the worst of architects from the best of bees', he writes, 'is that architects erect a structure in the imagination before realizing it in material form' (quoted in Harvey, 2000: 159).

Particularly relevant to understanding the significance that fantasy and imagination hold for neuromantic architecture, especially as manifest in 'paper-based' or 'fantastic' architectural schemes, is Robert Harbison's (1992) book *The Built, the Unbuilt and the Unbuildable: In Pursuit of Architectural Meaning*. In this text, Harbison argues that it is towards the periphery of architecture—the 'outer fields', as he puts it—that one is to find the 'centre': 'architectural meaning' (7–11). By this he means that architectural meaning resides in the periphery of, *faute de mieux*, 'terrestrial' architecture: gardens, ruins, monuments, and, most importantly, speculative designs. Architectural meaning, then, at least as Harbison conceives of it, is located in those projects that exhibit 'form freed from function'—such as 'fantastic' or 'paper-based' architecture (161–178).

This is significant in light of Novak's (1992) influential early account of neuromantic architecture, which he outlines in his essay, 'Liquid Architectures in Cyberspace'. In this essay, Novak places heavy emphasis on 'fantastic' architecture as an antecedent to neuromantic architecture, revisiting the great speculative and unrealised design proposals of architectural history, like those by such diverse architects as Castaglioni, Finsterlin, Fuller, Boullée, Gaudí, Le Corbusier, and Wright. Novak gives this work the collective tag of 'visionary' architecture. These schemes are so labelled because, for Novak (and other writers like him), they represent the precursors to cyberspace architecture—an architecture where form is freed from function, architecture that is (allegedly) outside the bounds of physics, 'pure' architecture, an architecture of the imagination. 'Visionary' (fantastic, paper-based, unbuilt) buildings are testimony to the fact that so much of architecture is in fact, like the classical *ars memoria*

(Yates, 1996), a mental activity, unfolding in the labyrinthine spaces of the imagination. And it is these paper-based 'visionary' buildings, these spaces of the imagination rather than built buildings, which hold most promise for Novak and others.

Moreover, for Novak in particular, the transformative power of the imagination is tied to a particular logic—the logic of poetic language and poetic thinking. He writes:

> Cyberspace is a habitat for the imagination. Our interaction with computers so far has primarily been one of clear, linear thinking. Poetic thinking is of an entirely different order. (Novak, 1992: 225–226)

Novak's interest in poetic language, as with his interest in 'visionary' ('fantastic') architecture, is that both are seen to be unfixed and fluid, both are spaces of possibility: 'Poetry is liquid language' (229); 'Poetic language is language in the process of making' (242).

The neuromantic architectural fascination with the transformative (and transgressive) possibilities of poetic language, fantasy, and imagination can be seen as a firmly Romantic sensibility, at least in the sense that Niall Lucy (2000) describes it, as an attempt at 'presenting the unpresentable'. Referring to the Romantic struggle with the ultimate unreproducibility of imaginative images and forms ('the limits of representation'), Lucy suggests that 'the impossibility of their presentation can be understood (thought and felt) only in their futile struggle to present them' (7).

But as Lucy is writing specifically in relation to postmodern literary theory, how does this pertain to neuromantic architecture? One such way is as a desire to transcend the limitations of structural engineering, construction, and physics. Freed from the usual constraints on architecture, at least in theory, neuromantic architecture would seem to present the possibility of finally realising this dream in 'built' form (apostrophised because it is *built* in the virtual space of computer graphics).

An alternative, admittedly paradoxical reading is that it is not so much the ability to finally 'realise' unbuildable buildings that makes cyberspace so fascinating to architects. Rather, the appeal lies in the extent to which designing and 'building' in cyberspace becomes a continuation of the Romantic struggle with the *limits* of representation and, perhaps, the ultimate unreproducibility of imaginative images and forms. This is evident in Novak's comment on the 'unpresentability' of his own notion of 'archimusic': 'while we can surely imagine such an art form, we have had no way to actually construct and inhabit the spatiotemporal edifices of that imagination' (Novak, 1995: 44). At least, not yet. Thus, to borrow the words of one architectural critic (writing in a different context), architecture—*and especially* 'neuromantic' or cyberspace architecture—'turns out to be a tantalic objective: the greater the efforts to achieve it, the more elusive it becomes' (Polano, 1988: 87).

This double concern—faith in the transformative powers of the imagination and pursuit of the limits of representation—points towards what is perhaps most productive about this reading of 'neuromanticism'. Despite this apparent unpresentability, the effort expended towards pursuing it can be worthwhile. It is through striving towards presentability that constructive opportunities emerge for understanding the wider creative potential of teletechnologies.

Novak discerns such constructive possibilities in poetic language. And, in his own (admittedly very different) way, so did Martin Heidegger (1971a) before him. For Heidegger, as Lucy (1997: 201) puts it, 'art is not simply a category of truth [. . .] rather it is what "unconceals" the truth of truth or brings truth to light', with art considered as a form of *poesis*—a way of 'bringing forth' or 'revealing'. Thus, in order to realise such 'unconcealment', it has been suggested that 'we must think poetically' (201).

Such an approach is useful for thinking about cyberspace and its architectures, in the sense that poetic thinking 'allows a non-utilitarian and non-objective truth about things to be unconcealed' (201). Novak recognises the value that Heidegger's philosophical investigations into the essence of art holds for cyberspace and its architectural form:

> Heidegger anticipates what we are presently witnessing: the reconvergence of art, science, and technology in techne, 'a single, manifold revealing' of the ways in which we know the world that is neither merely instrumentalized nor merely aestheticized, but that brings our present approximations of the true into the range of lived experience through questioning. (Novak, 1996)

The point I am making here, which differs somewhat from Novak's, is that such 'questioning' or 'unconcealment' brought about through poetic thinking—and the reconvergence of art, science, and technology—can be fruitful for gaining deeper insight into the opportunities and implications of teletechnologies, the virtual and virtual space, and how these things intertwine with questions of place and community.

Yet, it is important to proceed slowly, and pause to reflect on possible barriers to this process. In an essay included in the second of the two *Architectural Design* special profiles on 'architects in cyberspace', Paul Virilio offers the following pertinent *aperçu*:

> Each time we conceive an embryonic thought in the field of new technology, let us try without delay—with the lessons of the 20th century in mind—to think of the negativity hidden within it, which is already there. Let us try to approach it as engineers approach an accident. While working to avoid it, they are also working to achieve something positive. (Virilio, 1998: 61)

To borrow Heidegger's words, 'the realm in which the dialogue between poetry and thinking goes on can be discovered, reached, and explored in thought only slowly' (1971b: 98). It is necessary to be aware of some of the blockages or hindrances to this dialogue.

Issues Attending Neuromantic Architectural Theory

Some of these 'hindrances' and broader issues attending neuromantic architectural theory include: the framing of neuromantic architecture as radically 'new' in the way that it challenges existing architectural ideas and processes; its reliance on complex science; and its lack of full engagement with issues of place and community.

The first of these issues can be introduced by considering the way in which technology is celebrated in the architectural literature on cyberspace. In an essay addressing the task of thinking through the arrival, reception, and implications of 'new' technologies, Niall Lucy warns against what he terms 'autogenesis'. By this Lucy means that, 'one popular understanding of technology is that it brings about change at the expense of continuity; that it brings in the new only on the basis of an act of forgetting—forgetting the past', regardless of how this past is conceived (Lucy, 2001: 36; see also, Tofts and McKeich, 1998; Bolter and Grusin, 1999). Lucy's point is that 'technology is inseparable from human history—that is how we have recorded it, passed it on, kept it going. That's were we need to begin' (Lucy, 2001: 40). In the literature on cyberspace architecture, the tendency towards 'autogenesis' is most evident in the persistent claim that the technologies of networked computing (cyberspace) force significant revision of existing architectural 'construction-conceptions'.

Such is the commitment to detailing what is supposedly 'new' about these technologies and 'virtual' spaces, that challenges to these same 'construction-conceptions' from within the existing history of architectural theory and practice are often overlooked and pass unremarked on in this literature. There are in fact many potential points where the individual claims made in the writing of cyberspace architecture (documented in the first part of this chapter) have been pre-empted by other, earlier architects and architectural theorists. Further examination of cyberspace architecture in the context of historical precedents is a potentially fruitful endeavour for future research, as this comparative work is beyond the scope of the present book.

In addition to the achievements of architectural history, however, an even more fundamental challenge to these claims for radicalness emerges from a critique of the basic framing of cyberspace architecture as in conflict with architecture as otherwise understood and practised. Put very simply, in binary oppositional terms, for the former to be seen as 'radical', the latter has to be understood as staid and conventional. This involves a double assumption. First, that there is only architecture (singular), as opposed to architecture*s* (plural). Secondly, that this singular architecture is in fact 'Architecture'—that is to say, something that is monolithic, unchanging, and intractable. To challenge both assumptions in this too neat distinction 'menaces' or 'unsettles' the claims made for cyberspace. Not only does this invoke once again the poststructuralist idea that a 'repressed' term in an apparent binary opposition permits the possibility of the 'dominant' term, because both terms carry elements of each other, but it also points to the need to more fully historicise the apparent novelties of new technologies, including 'cyberspace'. This is important in that it would clarify in more detail what can and cannot legitimately be substantiated as 'radical' or truly 'new' in respect to the technologies and 'spaces' of networked computing.

Even the format of the *Architectural Design* journal itself works to undermine these claims to radicalness. The two themed issues of this journal on 'Architects in Cyberspace', which formed a key resource for my examinations of cyberspace architecture, include theoretical essays and, following these, a selection of 'supporting' 'visual essays' showcasing recent works by various (usually established) architects. But, if we are to look at the *Architectural Design* (*AD*) journal as a whole text, the juxtaposition of the written theoretical work and the illustrative work poses its own problems. These 'visual essays' at the back of each themed issue are intended to support the preceding written theoretical work. Yet, these illustrations relate to schemes for actual or proposed buildings, and, moreover, have often been developed independent of the theme and, in many cases, are incommensurate with the theme. This is evident in the case of the inclusion of Bernard Tschumi's 'Cité de l'Architecture' project in the first *AD* themed issue on 'Architects in Cyberspace' (1996). It is also evident in the inclusion of Arakawa and Gins's 'City of Reversible Destiny' in the second *AD* themed issue, 'Architects in Cyberspace II' (1998). Neither of these projects support, mirror, or even really engage with the sorts of ideas and concerns addressed in the theoretical writing on cyberspace and architecture that precede them in each journal issue. If anything, this juxtaposition of text and project images frames (or reinforces) these theoretical meditations on cyberspace as parallel or even supplementary to existing architectural design and practice, rather than mounting a challenge to them. In short, it re-establishes a theory/praxis divide.

Setting up cyberspace architecture as a radical challenge to existing architectural design processes also risks falling into the trap of a further dichotomy which establishes the 'virtual' as a distinct realm from the 'real'. (This in turn can give rise to, or fuels, yet another problematic distinction: the dichotomous conception of cyberspace, depending on one's point of view, as fundamentally either utopian or dystopian.) Many neuromantic architectural theorists are at pains to dispel the first of these misconceptions. For instance, McGuire (1995: 54) notes that cyberspace is *not* a distinct space from 'our own [space]', and Novak (1996) remarks that 'cyberspace, in its many forms, will not replace physical space'. (In this context it is perhaps fitting that neuromantic architectural theory was described earlier as a 'complex', as one dictionary meaning of the word complex, derived from mathematics, defines it as 'containing real and imaginary parts' (Hughes, Michell and Ramson, 1992: 224).) Despite such assurances, there is still lingering concern (if not widespread conviction) by other critics that just such a polarisation persists (Boyer, 1996; Franck, 1995, 1998; Pawley, 1998). The shadow of these characterisations remain, and there is a need to dismantle this way of thinking and develop fuller and more nuanced accounts of the full range of our social and spatial/place-based interactions with teletechnologies—a task that is taken up in more detail in the final chapter of this book.

Even here in this chapter, one difficulty associated with developing a picture of cyberspace architecture as a return to elements of Romanticism is that it risks the familiar split that sets the 'imagination' in opposition to 'reason'. It is an erroneous division, yet it has long influenced debate on literary Romanticism (Bowra, 1963), as it has writing on architectural Romanticism (see for example, Capon, 1999a, 1999b; Hellman, 1986: 128–129). This is a slip that I have consciously avoided in the present analysis.

The second key issue attending neuromantic architectural theory concerns its strong guiding interest in complexity theory. Doreen Massey refers to this in broad terms as a 'reliance on science'. Architectural interest in these influences is perhaps not surprising given Massey's (2005: 127) assertion that, 'in the most general of terms the theory of complexity evokes "the spatial"'. Nevertheless, Massey details a number of potential hazards that attend such dependence, which bear explanation here.

Massey begins by cautioning against 'unreflective acceptance' of scientific theories by 'those postmodern writers in social science and humanities [which should include architects] who today rest their case with [. . .] enthusiasm on "the new sciences"' (2005: 73).

This is not to say that Massey is necessarily antagonistic towards complex science. On the contrary, in a later passage she endorses Nigel Thrift's (1999) suggestion that 'complexity theory might be seen as one of the harbingers of something more, the emergence of a structure of feeling in Euro-American societies which frames the world as complex, irreducible, anti-closural and, in so doing, is producing a much greater sense of openness and possibility about the future' (quoted in Massey, 2005: 126). For this reason, Massey suggests that ideas of complexity may in fact be part of a 'wider *Zeitgeist*' (126). Accordingly, she argues:

> It may, then, be more appropriate to interpret references to complexity theory, even when [. . .] they appeal quite explicitly to a natural science as a legitimising ground for their argument, rather as particular elements in a wider and multiply-interconnected structure of intelligibility

which is emerging as common to the age, as least in Western countries. (Massey, 2005: 127)

In the context of the literature on architecture and cyberspace, the idea that complexity enables a greater sense of openness and possibility is particularly strong, and perhaps architectural interest in these ideas is merely reflective of wider societal interest in the same ideas (see, for example, Urry, 2005: 1). This particular take on complexity could prove quite useful insofar as it can shed light on architectural interest in complex science and how this interest fits within broader societal 'structures of intelligibility'.

Yet, and despite this, Massey is quick to qualify her above point by adding:

we are still duty-bound to [. . .] specify, each in our own field of study, just what we mean by hailing this general reference into our own particular area, and just what work it does, [and] upon what issues it gives us more effective purchase. (2005: 127)

This is where the architectural appeal to these theories of complexity needs to be treated with caution. For, on the one hand, it is possible that these theories could prompt renewed consideration of the process of (and the 'construction-conceptions' involved in) architectural design, as well as offering an expanded suite of tools to the architect-designer. On the other hand, problematic tendencies attend neuromantic architectural theory's engagement with complexity theory. There is the tendency to appeal quite explicitly to these theories as a legitimising ground for these theorists' arguments. There is also a tendency implicit in this particular strand of architectural writing to use this theory in a very crudely dichotomous way, where 'chaos' and 'complexity' are taken to be indubitably good, and any form of 'order' is unequivocally bad. This is similar to Massey's argument (and that of other critics) that '"previous bodies of scientific theory" were in fact *on their own readings* precisely abstracting *from* the historical messiness the reassuringly stable (for them "spatial") eternal truths' (128—original emphasis). However, the same 'oppositional' form of order—order as disorder—is repeated in architectural writing on cyberspace. What seems to pass unnoticed in this writing is that, by repeating this process, 'chaos' and 'complexity theory' become in neuromantic architectural theory no less an organising system than that which these theorists seek to oppose. What is unfortunate about this approach is that it forecloses on the more productive spatial possibilities which might be available through and beyond complex science (such as has been discussed by Thrift and Massey).

The third and final issue attending neuromantic architectural theory concerns the lack of any sustained engagement with ideas of place or community in the architectural literature on cyberspace examined in this chapter.

In a brief section on architecture in his book, *The Philosophy of Art*, the Romantic philosopher Friedrich von Schelling writes, 'There are certain kinds of architecture where need and utility fall completely by the wayside, and its works are themselves the expression of absolute ideas that are independent of need' (1989: 165). This seems to be the precise sense in which neuromantic architecture is framed and understood. The inevitable result is that questions of architectural function and occupation, the social, and place (in its fullest sense), are all (a)voided in the architectural discourse on cyberspace.

The lack of engagement with place is a somewhat surprising omission given the close interconnection between space and place, and given the argument that 'place cannot be divorced from space' (Malpas, 1999: 42). It is also a surprising omission given the centrality of metaphors of place in other writings on the virtual, particularly 'virtual community' (see Chapter 3).

Or is this omission really so surprising? When contrasted with the crucial role of place in the literature on 'virtual community', the absence of place in the literature on cyberspace architecture only serves to highlight Malpas's argument that 'it is within the structure of place that the very possibility of the social arises' (36). In this context, it is perhaps not that surprising an omission after all, especially given cyberspace architectural theory's belief in architecture as 'the expression of absolute ideas' and, because of this, the ambivalence (and at times, outright hostility) that is evident in this theory towards the social and the human subject/human body. Indeed, what this lack of engagement might suggest is that architectural theories of cyberspace constitute a return to elements of Romanticism, but that this return is skewed in particular directions—namely, towards a celebration of transcendence through technology at the expense of other concerns, including place and the socio-political.

THE PROMISE AND THE PERIL OF NEUROMANTIC ARCHITECTURAL THEORY

In the preceding chapter's examination of early architectural computing, the computer was approached as an aspect of a 'prehistory' of computers as a teletechnology. During this formative period, architectural theorists and designers understood the computer as a powerful computational tool for architectural use. For some, such as Yona Friedman and Nicholas Negroponte, it even had the potential to supplant the architect in certain key roles, particularly problem resolution. However, this problem resolution was still very much divorced from consideration—or at least, *conscious* consideration—of the computer's representational possibilities. By the 1990s, architectural theorists and designers had embraced the computer as a teletechnology—a technology of distance. Moreover, and in contrast to its earlier usage, the pre-eminent role of computers (and supporting computer-aided design software and other modelling languages) had shifted to that of a representational or form-generating technology. Problem-resolution had given way to form generation. Or, to put this slightly differently, the 'problem' facing architects in the digital age is taken to be 'Architecture' itself (especially its suite of existing 'construction-conceptions').

What remains consistent between early and more recent architectural computing is a strong and guiding interest in complex science. As noted in the previous chapter, scientific thought on complex systems (especially General Systems Theory and cybernetics) had a significant influence on experiments in early architectural computing. Two or three decades later, this influence continues (see Hunt, 1998; Rahim, 2000). Yet, a key issue attending this continuing 'reliance on science' (as Massey terms it) is the need to be clear about the 'work' that these theories perform within architectural discourses on new technologies, particularly in legitimating arguments about their transgressive potential.

Also consistent with early architectural computing is the failure of neuromantic architectural theory to engage in any meaningful way with the idea of 'place'—particularly in the sense that this term can be understood to involve the very structure of human activity and experience. Indeed, this lack of engagement with ideas of place, if anything, is even more pronounced than in earlier writing on architectural computing, or populist and critical writing on computer-mediated communications and 'community'.

Likewise, any engagement with or sense of 'community' or sociality or even embodied existence in cyberspace is altogether absent from neuromantic architectural theory. This particular lacuna is perhaps not that surprising given that many of the more contentious claims made about disembodiment that have been proposed in techno-boosterist writing on the Internet form an important basis for neuromantic architectural theory.

In a 1998 essay, Marcos Novak claims that to be involved in architectural design in cyberspace one 'must be assertive and deeply involved in languages and techniques particular to virtuality' (1998: 22). He continues:

> It is this requirement that inevitably leads to algorithmic design and the exploration of mathematical spaces outside our normal, Euclidean/Cartesian/Newtonian experience, beyond the scale and dimensionality of bodyspace. (Novak, 1998: 22)

Even so, Novak still maintains that 'such sojourns are not attempts to avoid the real, as the fearful have tried to trope them' (22). On the contrary, he asserts, 'they are attempts to engage the real of virtuality in ever more rigorous ways' (22). As Novak (1997: 397–398) writes elsewhere, the virtual 'will naturally lead us to the inhabitation of increasingly abstract spaces, regardless of how familiar and comforting we choose to make the surface appearance of the interface between here and there'.

This passage by Novak points towards both the promise and the peril of neuromantic architectural theory. On the positive side, there is a great deal that is productive and encouraging about the emphasis in this theory of the transformative potential of the imagination—particularly for developing, as Massey (2005: 128) puts it, a 'revolutionised imagination of space'. On the less positive side, these theorists fail to extend their spatial re-imaginings to questions of the socio-spatial (and to what Massey refers to as the 'throwntogetherness of place' and the 'event of place') and, because of this, to questions of the socio-political (2005: 140–141). In other words, the neuromantic architectural theorist's vision of the virtual is arguably an impoverished vision—spatially, socially, and politically. In Lefebvrean (2000) terms, what is lacking from neuromantic architectural theory is any real engagement with the full complexities of 'social space' (see also, Forty, 2004: 270–275); for Lefebvre (2000), this is the 'space' within which the cultural life of societies takes place.

In lieu of these socio-spatial and socio-political questions, neuromantic architectural theorists are, to borrow Scott McQuire's words, 'inevitably drawn towards the recapitulation of fictions of individual mastery and transcendence via technology, without registering the extent to which technological transformation has altered the social parameters within which individual identity is embedded' (2002: 177).

Thus, in order to develop the possibilities discernible in the emergent field of neuromantic architecture—specifically the potential that can be sourced in poetic thinking and the transformative powers of the imagination—it is necessary to wrest these ideas from the language of 'cyberhype' (Wark, 1997), to distance them from the techno-boosterist celebration of cyberspace as a 'dematerialised', 'disembodied' realm unfettered by the constraints of physics, geography, and the flesh. A further challenge is to reorient these computer-mediated spatial re-imaginings to address questions of place, and the socio-political ('community'). These challenges need to be addressed in order to begin to develop a more complete and more detailed picture of the interactions between teletechnologies, place, and the social/communal.

The previous two chapters have focused specifically on architectural engagement with computing technologies. In the next chapter, this interest in space and technology use turns to more socially situated technology use, both within and beyond a particular place—that of the domestic home—and, in contrast, across urban places, through a focus on mobile phones and the pedestrian negotiation of cities.

6   Domesticating Technology, Mobilising Place

'Why is Le Corbusier's famous remark about a house as "machine for living" so often quoted at us? [. . .] It is the abrasiveness of the machine as an image which remains when it is applied to that most comfortable of traditional values, the home.'

—Peter Cook (1967: 18)

'Home is no longer just one place. It is locations.'

—bell hooks (1990: 148)

The previous chapters have charted different contexts in which the varying interactions between place, community, and (tele)technology are played out. In some contexts, such as scholarship on computer-mediated communication from the 1980s and 1990s, place is predominantly explored as a restricted metaphor, while community, as seen in Chapter 1, emerged at the same time as a dominant metaphor. In other contexts, such as architectural writing on cyberspace, notions of place and community have minimal representation.

In the light of these examinations, the present chapter looks closely at two cases where the relationships of place, community, and teletechnology are more closely interconnected. It asks, how do place, community, and teletechnology function in the highly personalised—even ubiquitous—context of the domestic environment, the 'home' as a particular 'place' and 'community'? Do teletechnologies modify existing conceptions and experiences of home, and if so, how? And, following from this, what happens when ideas of home are 'displaced' or altered by mobile technologies and their use? Does place have any relevance in the context of networked mobility, or is it rendered obsolete? And, by contrast, what happens to common conceptions of place as a 'proper, stable, and distinct location' (Morse, 1990: 195) as a result of mobile practices?

Therefore, the aim of this chapter is, on the one hand, to explore ideas of 'home', and the impacts of new technologies on the home, as understood in the literature in this field. It does this by considering notions of 'home' as a key architectural and socio-cultural site, and the consumption of technologies in this context. This context includes the consumption of convergent or 'smart house' technologies at one end of the spectrum, and the incorporation and use—the 'domestication'—within the home of individual information and communication technologies, such as the computer, at the other end of the spectrum.

The further aim of this chapter, on the other hand, is to examine a key area of contemporary teletechnology use: the rise of networked—especially telephonic—mobility, and the far-reaching impact this has on the way space and place (especially urban space and place) are experienced and understood.

This dual investigation is valuable for a range of reasons. The home is one of the most personalised and important sites of European and Anglo-American 'place' and 'community' (Easthope, 2004), and is also potentially a highly technologically-mediated environment. Meanwhile, mobile technologies are one of the key and expanding contexts in which teletechnologies are employed and consumed. This dual focus is also valuable in that it shifts the focus from broad-scale planning issues, and at times abstract design theories, to the context of the everyday or quotidian. It also brings the *user* into clearer focus, rather than the *designer*.

To preface these deliberations, and in order to understand the full impact and implications of teletechnologies within and beyond the domestic home, it is important to first consider how the notion of 'home' is understood and what is at stake in these understandings.

THE CONCEPT OF 'HOME'

The concept of a 'house' generally refers to a 'physical unit that defines and delimits space for the members of a household. It provides shelter and protection for domestic activities' (Lawrence, 1987: 155). According to Gwendolyn Wright (1991: 215), the English word 'house', like the German word *Heim*, is defined as 'offering protection and familiarity', and carries 'allusions to the walls of the Old English *hus*, where goods were safely stored and husbanded'.

'Home', on the other hand, is a more elusive and complex notion that extends beyond structure and the need for enclosure. In its domestic sense:

> A home is more than just 'a territorial core' [. . .] and not just 'an ordering principle in space' [. . .], but a complex entity that defines and is defined by cultural, sociodemographic, psychological, political, and economic factors. (Lawrence, 1987: 155)

In other words, not only is the domestic home 'a place of familial and moral value' (Tabor, 1998: 221), 'offering protection and familiarity' and with its own 'moral economy' (Silverstone, Hirsch, and Morley, 1992: 18–19), but 'it [also] has more psychological resonance and social meaning' (Lawrence, 1987: 155; Chapman, 2001).[1] In short, 'unlike the house, [. . .]

home is a subjective construct, a metaphor of the self and the body': 'A house identified as the self is called "home"' (Tabor, 1998: 219 & 218).

Yet, there are also other, wider understandings of 'home'.

At a geopolitical scale, and in 'a world of expanding horizons and dissolving boundaries', 'home' becomes a politically and culturally charged notion with almost talismanic appeal (Morley and Robins, 1995: 86). For example, it was invoked by Mikhail Gorbachev, who was the final head of state of the former USSR, in his call for a 'common European home' (87). It emerges in the desire for 'another kind of "homely" belonging [that is] rooted in the Heimat of regions and small nations' and 'as an alternative to continental Europeanism and to nation statism'—a desire expressed by Basque separatists and many other cultural groupings (89). And, it is perhaps most poignantly felt in the tough border protection policies currently favoured by many developed countries, particularly Australia during the years of the Howard government. Spanning all of these:

> There is an increasingly felt need for 'some expressive relationship to the past' and for attachment to particular territorial locations as 'nodes of association and continuity, bounding cultures and communities'. There is a desire to be 'at home' in the new and disorientating global space. (Morley and Robins, 1995: 87)

Teletechnologies are implicated in this process. For example, in commenting at length on the defamiliarising effects of telecommunications technologies and the forces of globalisation, Derrida observes that they lead to a growing and renewed desire for the 'home'—in both its domestic sense and in a potentially more threatening nationalist sense:

> The global and the dominant effect of television, the telephone, the fax machine, satellites, the accelerated circulation of images, discourse, etc., is that the *here-and-now* becomes uncertain, without guarantee: anchoredness, rootedness, the *at-home* [*le chez-soi*] are radically contested. Dislodged. This is nothing new. It has always been this way. The *at-home* has always been tormented by the other, the guest, by the threat of expropriation. It is constituted only in this threat. But today, we are witnessing such a radical expropriation, deterritorialization, delocalization, dissociation of the political and the local, of the national, of the nation-state and the local, that the response, or rather the reaction, becomes: 'I want to be *at home*, I want finally to be at home, with my own, close to my friends and family.' [. . .] The more powerful and violent the technological expropriation, the delocalization, the more powerful, naturally, the recourse to the at-home, the return toward home. (Derrida and Stiegler, 2002: 79–80)

In a somewhat different but still connected sense, Angelika Bammer defines 'home' as 'always existing in the virtual space between loss and recuperation [. . .] the imaginary point where here and there—where we are and where we come from—are momentarily grounded' (1992: ix). In this sense, '"home" may refer to a deeply familiar or foreign place, or it may be no more than a passing point of reference' (vii). What this generates is 'a sense of home as always existing in the virtual space between loss and recuperation' (ix).

These conceptions of 'home' appear far removed from how 'home' is understood in a domestic sense. Yet, as will emerge in what is to follow, these are by no means dissonant understandings. Indeed, addressing the challenges and possibilities that such broader approaches to thinking 'home' pose—the question of 'what [. . .] to do with "home"', as Bammer (1992: xi) frames it—will prove important in making sense of the wider implications and intersections of domestic teletechnologies with understandings of place and community. These are, however, issues that are to be set aside until later. For the present, discussion will be limited to 'home' in its more precise—yet still 'elusive'—domestic sense, where there are two key points to be made.

The first of these is that 'home' is often understood to function both as a kind of cradle for wider community—especially, as Wright (1991: 221) notes, by conservatives and progressive reformers—as well as a form of 'community' in its own right. In respect to the second of these, 'home' has been described as an 'embryonic community' (Douglas, 1991: 288) and a 'virtual community' (297), with 'virtual' in this instance taken to mean 'independent' or 'autonomous' (292). This is an important conceptualisation given that the notion of 'community'—especially as it intersects with understandings of place and teletechnology use—is a core concern of this book. It is also valuable to bear this understanding in mind in light of the discussion to follow below concerning the impacts of information and communication technologies on the home.

The second point is the idea of 'home' and the 'domestic ideal' as a 'utopian community, [. . .] a retreat from the world' (Hareven, 1991: 262–263).[2] Hareven writes:

> Following the removal of the workplace from the home as a result of urbanization and industrialization, the household was recast as the family's private retreat, and home emerged as a new concept and existence. (1991: 259)

This second point follows on from the first insofar as the notion of home as an embryonic community to a certain extent requires privacy for its sustenance. Even so, as Hareven argues, this is a relatively recent development:

> The concept of the home as a private retreat first emerged in the lives of bourgeois families in eighteenth-century France and England, and in the United States among urban, middle-class families in the early part of the nineteenth century. Its development was closely linked to

the new ideals of domesticity and privacy that were associated with the characteristics of the modern family—a family that was child-centered, private, and in which the roles of husband and wife were segregated into public and domestic spheres, respectively. (Hareven, 1991: 258)

Nonetheless, it is one with lasting consequences for twentieth and twenty-first century understandings of urban space and culture.[3] Described here, too, are the seeds of the modern 'domestic ideal' of home ownership, especially as a 'retreat from the world [. . .] that had to be consciously designed and perfectly managed' (263; see also, Edwards, 2005).

The reasons for this emergence of 'home' as retreat are manifold, and predate but mirror many of the same factors that have influenced twentieth century planning thought from Jacobs (1972) and Mumford (1979) through to the Congress of the New Urbanism (Leccese and McCormick, 2000):

> The idealization of the home as a haven was a reaction to the anxiety provoked by rapid urbanization, resulting in the transformation of old neighbourhoods and the creation of new ones, the rapid influx of immigrants into urban areas, and the visible concentration of poverty in cities. (Hareven, 1991: 263)

Needless to say, the idea of home as a 'consciously designed and perfectly managed' retreat has experienced sustained and ongoing critical examination over several decades. The domestic home has been criticised as a 'locus of regressive nostalgia', to borrow Bammer's (1992: x) phrase; a stifling, conservative, and exclusionary institution (as Vilém Flusser (1990: 35) puts it, 'A wall is a tool to protect the inhabitant from what is different'); and a 'contested zone' (Smith, 1993: 104) marked by gender divisions and repressive and oppressive sexual politics sustained by unequal distributions of power and individual freedoms.

Feminism, especially early second wave feminism (including the likes of Simone de Beauvoir, Betty Friedan, Germaine Greer, Anne Oakley, and Anne Summers, among others), has been instrumental in developing and driving this critique. In an important recent review of this contribution, Lesley Johnson and Justine Lloyd (2004) acknowledge the lasting importance of this critique as well as the problems it poses for contemporary feminism. Johnson and Lloyd's project is to 'unsettle our received memories' and show how the home and especially 'the housewife role, far from silencing women, actually gave women as a group a speaking position, and a political voice' (2004: 156). They further argue the importance of the home as a site through which families create meaningful lives for themselves:[4]

> where this would seem to work best in terms of creating a robust sense of one's own individuality is when the home is open to the world rather

than closed in upon itself: the home as a source of values and strength to enable with and reflect upon the world, its social practices and structures. (Johnson and Lloyd, 2004: 156)

But, they add, because of this:

The question remains of how to continue this understanding of past changes to the domestic sphere in a way that takes account of the wide networks in which we operate, and the global circuits of exchange that now constitute our homes. (2004: 157)

It is not the task of the present discussion to respond to the challenges posed by this question. Suffice to say, however, that this quote alludes to the fact that contemporary teletechnologies and global information flows and networks pose significant further challenges to how we understand home (not to mention women's place in it).

Thus, having developed a fuller understanding of what is meant by and bundled up in the notion of 'home', the following sections examine how teletechnologies are impacting on the domestic home and its functions—especially as these impacts are said to be profound.

CHANGING INFORMATION AND COMMUNICATION TECHNOLOGY IN THE DOMESTIC HOME

Information and communication technologies (or teletechnologies), it is argued, constitute a 'telematic invasion of our literal and inner homes' (Tabor, 1998: 224):

These days a house consisting of a roof and four walls can only be found in a fairy tale. The earthquake called the communications revolution has reduced it to ruins. Material and immaterial cables have made an Emmenthal cheese of it [. . .]. We don't live in houses anymore, we hide in ruins through which blow the blizzards of communication. (Flusser, 1990: 36)

While not denying the significant impacts that teletechnologies have had on the home, it is worth prefacing a discussion of them, and tempering the hyperbole in the above statements, by questioning the myth of a hitherto secure and private domestic home on at least two levels.

The first is by recognising that our understanding of 'home' as enclosed and private is a comparatively recent one, and the domestic house/home was not always understood this way. For example, Hareven observes that, 'In preindustrial society there was a significant difference between the family's domicile—the household—and the *home* as it became idealized later in

Western European and American society' (1991: 255—original emphasis). Drawing on the work of Philippe Ariès, Hareven argues that, 'the family in preindustrial society was characterized by *sociability* rather than *privacy*' (256—original emphasis), and that, in contrast to its contemporary conception as a private retreat, in preindustrial society it was *inside* the household where the family conducted its work and public affairs:

> Households were teeming with various activities, and family members, even couples, could hardly retreat into privacy within the crowded household space. The head of the household's various business associates and other individuals actively involved in the family's economic and social activities were often present in the household. (Hareven, 1991: 256)

In keeping with these arrangements, even the internal 'spaces within the "big house" were not differentiated into family space and public space' (257). However, and as already noted earlier, this arrangement began to change from the late eighteenth century onwards,[5] where 'the new domestic style [for a nineteenth century middle class] required two types of privacy: privacy of the family from the community, and privacy of family members from each other *within* the home' (268—original emphasis).[6]

Philip Tabor (1998: 218) suggests that, 'as far as the idea of home is concerned, the home of the home is the Netherlands'. He writes that in 'the first three-quarters of the seventeenth-century [. . .] the Dutch Netherlands amassed an unprecedented and unrivalled accumulation of capital' through international trade and colonial expansion, 'and emptied their purses into domestic space' (218). Even so, Tabor argues that it was nevertheless a period marked by a profound unease regarding the separation of the domestic and the foreign, the inside and the outside—both at the national and domestic levels. These tensions, he suggests, are strongly evident in Flemish painting of the period.

A good visual illustration of the tensions between private and public and security and openness of which Tabor writes is the following image by the painter Pieter de Hooch, entitled *The Courtyard of a House in Delft* (1658) (see Figure 6.1). In this painting, these tensions are depicted in a very literal sense. All the elements in the composition of the image—the angled plank to the right of centre, the change in colour in the archway brickwork, with the lighter bricks closer to the centre of the frame, and the open window shutter to the left of frame—work to lead the eye from the inside courtyard (itself both an inside/outside, public/private space), through the arch and towards the open door at the front of the house. It is in many respects an awkwardly composed image that nevertheless still works to create a sense of latent tension and unease—an 'uncanny' feeling—in its depiction of everyday domestic life and its relationship to the spaces and events beyond the domestic home.

Secondly, the idea of the 'secure home' has, to a certain extent, always been something of a fiction, with the domestic home far more porous and open to outside influence than is generally acknowledged.[7] As Doreen Massey writes, 'that place called home was never an unmediated experience'

Figure 6.1 This painting provides a strong visual illustration of the tensions between private and public and between security and openness which characterise the domestic home in history. This image also serves as a visual metaphor for the intensification of these tensions in the modern domestic home as a result of the impacts of teletechnologies.
Source: Pieter de Hooch, *The Courtyard of a House in Delft* (1658). Oil on canvas, 73.5 x 60 cm. National Gallery, London. Reproduced with permission.

(1992a: 8). While Massey is writing of 'home' in a somewhat more expansive sense, this statement is nevertheless just as pertinent to understandings of 'home' in a domestic sense. For example, at a basic level, 'the home serves as a means of communication with oneself, between members of the same household, friends, and strangers' (Lawrence, 1987: 161). This is the case particularly with respect to material culture, as will be covered later in the chapter.[8]

The above observations are by no means intended to deny the far-reaching impacts of contemporary teletechnologies on 'home'. Indeed, it is fair to say that the notion of 'home' in the domestic sense has undergone profound changes as a result of teletechnologies (to which discussion will turn directly). Rather, the point here is to place this 'invasion' in historical perspective, and, indeed, to rethink it as something other than an 'invasion' or 'breaching' of domestic security and privacy. It is arguably more fruitful to conceive of teletechnologies as representing an *intensification* of the channels or flows of information and communication into and out of the domestic home, albeit an intensification that raises significant and new challenges for thinking 'home'. Perhaps it is useful to bear in mind the image by Pieter de Hooch, and to think of this image, and the tensions embodied in it, in two ways. First, as a visual reminder of the tensions structuring the domestic home through history. And, second, as a distillation of, or metaphor for, the intensification of these tensions as a result of the considerable impacts of information and communication technologies on the modern domestic home.

In his (1985) study, *No Sense of Place*, Joshua Meyrowitz writes about how, through radio, TV, and phone, the family home has become 'a less bounded and unique environment' (vii):

> Electronic media have increasingly encroached on the situations that take place in physically defined settings. More and more, the form of mediated communication has come to resemble the form of live face-to-face interaction. More and more, media make us 'direct' audiences to performances that are not physically present. (Meyrowitz, 1985: 7)

As a result, 'the media pose a whole set of control problems for the household, problems of regulation and of boundary maintenance' (Silverstone, Hirsch, and Morley, 1994: 20). As Silverstone and his colleagues observe:

> These [problems] are expressed generally in the regular cycle of moral panics around new media or new media content, but on an everyday level, in individual households, they are expressed through decisions to include and exclude media content and to regulate within the household who watches what and who listens to and plays with and uses what. (Silverstone, Hirsch, and Morley, 1994: 20)

Yet, electronic media not only encroach on and shape situations and behaviour, they are also increasingly *embedded* in and part of the very fabric of domestic life and the domestic house itself.

Mapping and making sense of both sets of transformations is the concern of the discussion to follow below, and which begins with an examination of the architectural and teletechnological phenomenon of the 'smart house'.

The 'Smart House'

Terms such as 'smart house', 'electronic house', and 'e-house' are used to describe an integrated system of 'technological convergence' within the architectural setting of the domestic house. Technological convergence in this instance refers to the interconnection of various household appliances, telecommunications equipment, environmental control devices, and security systems, all of which are fed into and controlled via a central processing unit (CPU). The common vision of this technology, at least as it has been presented in the popular press, is of an 'intelligent environment' that not only controls various environmental factors (such as lighting) and responds to security breaches, but also pre-empts user needs. As Nicholas Negroponte (1995: 213) put it, 'If your refrigerator notices that you are out of milk, it can ask your car to remind you to pick some up on your way home'.[9]

There are numerous historical precursors to the 'smart house' idea—including, to name a few, Charles Barry's 1839 London Reform Club with its 'largely invisible heating, ventilation and systems of mechanical communication buried within the walls' (Forty, 2004: 89), William Randolf Hurst's opulent 1925 California castle San Simeon with its multiple radios tuned to different frequencies and piped into his private suite (Gates, 1996: 248; Coffman, 2003), as well as Ant Farm's 1973 experimental *House of the Century* project in Texas (Lewallen and Seid, 2004); this last example will be returned to later in this chapter. However, the paradigmatic modern example of the 'smart house' is Bill Gates's much-publicised $US35-million residential complex near Seattle, Washington. The Gates's house is said to include, among many other technological features, a wall of video monitors to screen images from his own digitised collection of artworks and photographs that have been licensed from galleries all over the world (Gates, 1996; Madsen, 1996). But of all the domestic technologies, the one Gates spends most time detailing in his own written account of the project is the electronic pins visitors to the house are given upon entering. 'The electronic pin you wear,' he writes, 'will tell the house who and where you are, and the house will use this information to try to meet and even anticipate your needs—all as unobtrusively as possible.' (Gates, 1996: 251) This may involve re-adjusting environmental settings to visitors' preferred temperature and light levels, as well as catering to individual audiovisual tastes of each guest as they move from room to room.

The potential implications of technologies such as those featured in the Gates's house are interpreted in widely divergent ways. For instance, proponents of 'smart house' or 'convergent' technologies see in them both the opportunity for enhanced responsiveness to environmental considerations and greater energy efficiency (McDonough, 2003; Plunkett, 1994: 100), as well as convenience and increased choice through the tailoring to and accommodation of individual taste (Gates, 1996; Negroponte, 1995; Keenan, 1999; Barker, 1999; Plunkett, 1994; Eccleston, 2000; Howard, 1994: 100). Furthermore, and crucially, given the earlier discussion of home and privacy, these technologies are said to bring greater flexibility to the domestic home through computer connectivity with global communications and information systems. As one commentator writes of the combining of computer and communications applications:

> The home can take on a meaning and function far beyond that of the 'family retreat'. In essence, it can be an activity centre for a wide range of business, education and entertainment applications. The home will be a repository and disseminator of information, providing services and benefits accessible internally and externally—even into the future information superhighways. (Plunkett, 1994: 99)

However, for many critics this is precisely what is most troubling about smart house technologies and digital convergence. For instance, Kevin Robins and James Cornford (1990: 872) claim that what motivates schemes for 'electronic homes of the future' is a desire—whether conscious or not—to 'subordinate the whole of social life to the principles of efficiency and control'. In accordance with this view, Fiona Allon argues that:

> The mechanistic and rigid instrumentation of Fordism and Taylorism has been superseded by an altogether more 'natural' form of control: the technologies have become soft and flexible, the planning and discipline of form and space replaced by an emphasis on individual comfort and creative productivity. (Allon, 2001: 16)

The problem with this, according to Allon, is that 'in the public / private work-space of the home, the control of time (work-time / private time) rather than spatial separation becomes the site for the exploitation and ordering of bodies' (2001: 18). As other critics have noted, 'The idea behind the digital home network is to seamlessly interconnect appliances into the very texture of domestic space and routine, to such an extent that even when [we] are not at home, home is never far away and is always within the reach of our control' (Gye and Tofts, 2003). The potential exists for a sharply intensified version of the preindustrialised home: one which blends leisure and work (as documented by Hareven and Ariès and described above); one where 'the move is towards work that may be performed at any time (and when the occasion demands, all the time)' (Robins and Webster, cited in Allon, 2001: 18); and where 'factory, home, and market are integrated on a new scale' (Haraway cited in Allon, 2001: 18; see also, Spigel, 2005).[10]

These are rather bleak assessments. However, in overall terms, what these socio-technological developments point towards is a complication and further blurring of the boundaries between public and private, inside and outside, and particularly the boundaries between individuals and corporate interests. These boundaries, Allon argues, 'are indistinct and permeable, constantly crossed by the traffic of information flows, undermining completely the model of the private individual ensconced in an enclosed private space' (2001: 17). As a result of these flows, public and private divisions (and all their 'related dispositions', such as sociality and privacy, connectedness and seclusion) 'refer less to enclosed and defined spaces, "spheres", than to *modulations* of engagement (frequently with streams of information both local and global) and to infinitely variable conditions of interaction and control' (17–18—original emphasis).

These two themes—that of the modern domestic home as characterised by the 'traffic of information flows', and where public and private are negotiated via 'modulations of engagement'—are both important ideas. Together they provide a sound and productive model for accounting for the tensions that structure sociality and privacy, connectedness and seclusion, within the technologised home. These ideas are also particularly productive for approaching these same issues of sociality and privacy and their associated tensions in relation to computer-mediated communications and, as will be discussed later in this chapter, in relation to networked mobility and mobile phone use.

Allon's remarks on the 'traffic of information flows' also share much in common with Scott McQuire's work on telecommunications and home. In commenting on Gates' account of his own house and its myriad technological features, McQuire (1997b: 527) writes: 'reading this description [by Gates] gave me a sensation which could best be described as *uncanny*'.

The Technological Uncanny and the Modern Unhomely

The 'uncanny' proves a particularly useful notion for McQuire for approaching the issue of electronic communication technologies and home. Discussing the use of the German word *unheimlich* in Freud's essay 'The "Uncanny"' (1971), McQuire writes:

> *Unheimlich* is often translated as 'uncanny' but could be more literally rendered as 'unhomely', and I find this double sense most suggestive.

> For Freud, the sensation of the uncanny is not caused so much by what is strange or unfamiliar as by the known and the familiar being made strange. Uncanniness suggests a disturbed domesticity, the return of the familiar in an apparently unfamiliar form. (McQuire, 1997a: 684)

The categories that Freud deploys in exploring the idea of the 'uncanny', each of which is linked to the experience of ambivalence, are for McQuire 'peculiarly suited to exploring contemporary communication technologies, particularly their effects in rearranging bodies and presences, times and spaces, seemingly at will' (1997a: 684). McQuire coins the term the 'technological uncanny' to capture the ongoing relevance of Freud's meditations on the 'uncanny'/'unhomely' for thinking through the interconnections of communications technologies and 'home' (1997a: 691).

Philip Tabor pursues a similar line of thinking with his notion of the 'telematic uncanny'. This describes the sense that 'electronic media have partly eroded not only social boundaries which previously divided individuals and families from society as a whole, but also some boundaries of the self which previously defined individual identity' (Tabor, 1998: 224). Tabor's statement about the impact of teletechnologies on social boundaries constitutes one of the key arguments concerning the impact of these technologies on the fabric of the domestic home. His point about ontological erosion, however, is somewhat more contentious. Drawing on Freud's essay and its examination of the experience of ambivalence, particularly as it concerns animate and inanimate objects, Tabor builds a familiar argument regarding the potentially negative impact of telecommunications technologies on the human subject and identity:

> Those who spend a large proportion of their conscious life on the Net or navigating informatic space may be prey, if only fleetingly and unconsciously, to feelings that barriers of identity are dissolving between selfhood and otherhood; that the mechanisms of resistance and causality, which had assured us we were separate from the outside world, no longer operate; that we float in a space outside the self. (Tabor, 1998: 223–224)

Up until the mid-to-late 1990s, such proclamations circulated quite widely and were received sympathetically by a number of critics. However, as discussed earlier in this book, it is a position which carries far less currency a decade or so hence given the growing body of literature that argues virtual technologies to be thoroughly embodied.

McQuire's engagement with Freud's essay, on the other hand, arguably opens up a more productive path in its consideration of the technological uncanny, first, by drawing connections between home in a domestic sense and home in a broader sense, and, second, by emphasising the 'ambivalence' of what Allon calls the 'traffic of information flows' (which Tabor argues leads to an erosion of social boundaries). In respect to the first of these two points, McQuire writes:

> The globalization of telecommunications flow goes hand in hand with the reorganization of the space of domestic life, including the micro-politics of the family. In fact, the most significant change is that where these fronts or frontiers—domestic, local, regional, transnational— were once distinct, or, at least, were believed to be, they now seem to be irreducibly imbricated in one another. (McQuire, 1997b: 528–529)

Elsewhere he adds, 'the theme of the uncanny also serves to join the current instability affecting "house and home" to more general questions of identity, cultural displacement, exile and homelessness' (1997a: 684–685). In a 1996 interview, McQuire observes:

> When we ask 'Where are we at home in the present? Who are our neighbours? What is the role of the stranger?', it seems to me that new media technologies produce similar effects to the cumulative impact of migration. (McQuire, quoted in Madsen, 1996: 48)

These important observations have been made in slightly different but no less forceful ways by other critics (Morley, 2000; Morley and Robins, 1995; Massey, 2005; Urry, 2000b). Some aspects of this relationship between multiple meanings of home will emerge in the discussion of networked mobility later in this chapter; they will also be taken up in more detail in the final chapter.[11]

The second significant point—which circles back to and reiterates Allon's remark about the 'traffic of information flows'—is McQuire's (2003: 103) claim that the contemporary, technologised home is being 'reposition[ed] as an interactive media centre'. For McQuire, to conceive of the home as an 'interactive node' that is permanently connected to vast information flows, 'radically alters the division and dynamics of public and private space, *including the notion of community* and the mode of linkage envisaged between its citizens' (1997a: 689—emphasis added).

There are two architecture-art projects worth noting briefly at this point which explore and exploit these inside/outside and public/private tensions and transformations in quite interesting and productive ways. The first is the MoMA-staged *The Un-Private House* exhibition (1999) and, particularly, Mojgan and Gisue Hariri's 'Digital House' proposal, with its interior and exterior walls consisting of 'smart skins' for the transmission of information (see Norwich, 1999; Spigel, 2001). The latter scheme resembles a cross between the Gates house with its bank of video monitors and the all-glass Panopticon dwelling depicted in Yevgeny Zamyatin's dystopian novel, *We* (Zamyatin, 1972). It is a proposal which explores the implications of ongoing changes in building technologies (or what is sometimes referred to as 'technology transfer'),[12] domestic consumption of teletechnologies, and the wider applications and implications of both (including the potential for surveillance). The second is the hugely ambitious 'Media House Project'—a joint venture involving the Metapolis Group from Barcelona, the Fundació Politècnica de Catalunya, and MIT Media Lab (Guallart, 2004). In their own words, this project sought to move 'information technologies beyond that of computers and integrate them into everyday life, literally looking to build computers from the components of buildings [and vice versa]' (Guallart, 2004: 14). The aim (and eventual outcome) was to incorporate in one habitable environment 'the physical structure, the electrical network, and the data network' (15). The idea was that this would enable a 'dynamic and configurable link between the entities (people, objects, space, limits, networks and contents)' both within and beyond this environment (15).

These two projects (and the earlier passage by McQuire) emphasise the understanding that contemporary teletechnologies not only challenge the myth of domestic privacy and the 'taken-for-granted function of the house' to use Cedric Price's words (more on this later), but also the conception of the home and family as an embryonic or 'virtual' (independent) community. As Deleuze (1992: 4) writes, 'the family is an "interior" in crisis like all other interiors', and one that is experiencing ongoing 'revolution' (VanEvery, 1999). This is not to say that home will cease to function as a private retreat and as a cradle of community; as Kim Dovey (after Bachelard) suggests, 'there are some unchanging things about home[;] it tends to be a stabiliser for people's identities' (quoted in Eccleston, 2000: 24). The point, rather, is that how 'home'—in all senses, but particularly the domestic—will continue to perform this 'stabilising' function is under significant (re)negotiation.

It is also worth emphasising that both of the above points by McQuire dovetail with David Morley's (2003: 3) belief that 'articulation of the domestic household into the "symbolic family" of the nation (or wider group) can best be understood by focusing on the role of media and communications technologies' within the home. This claim underscores the importance of what Clive Edwards (2005: 3) refers to as the 'material culture approach' to studying the consumption and uses of ICTs within the domestic home. As Marcel O'Gorman (2004: 36) points out, research in communications theory and cultural studies has 'gathered important data on how communications devices impact the domestic scene'. The focus of this discussion now shifts to these studies of the consumption of domestic technologies.

DOMESTIC CONSUMPTION OF TELETECHNOLOGIES

The available literature on the consumption of teletechnologies within the domestic home is substantial, and an exhaustive account of this research is not possible here. Rather, the present section surveys and summarises some of the key insights of research into the domestic consumption of teletechnologies (or ICTs), particularly home computers. Much of this work builds on the foundations of previous research on television and other earlier domestic technologies.[13] This research reveals that domestic teletechnology consumption processes are complex, shaping both the meanings and uses of domestic space.[14] The value of this material for the present discussion is that it challenges prevailing understandings of the home as a closed space, and strengthens the arguments made by Allon and McQuire regarding the interpenetration of the domestic home by teletechnologies. This material also provides a useful frame for understanding the complex interactions between home and practices of networked mobility, which will be developed later in the chapter.

A useful place to start is with Elaine Lally's (2002) detailed empirical study of computers in the domestic home. From this study, in which 95 individuals were interviewed about the consumption and use of their home computers, a number of key and interconnected insights emerged.

To begin with is the discovery that the processes contributing to the selection and acquisition of computers for home use are multilayered and at times difficult, and are often fuelled by 'a pervasive sense of the inevitability of the deepening penetration of information technology into everyday life, both within and outside the domestic sphere' (Lally, 2002: 48–50; Lahtinen, 2000), which is coupled with a sense of the importance of keeping up with this technology or risking being left behind.

This is only the beginning. Once an individual teletechnology (or ICT)—such as a new computer—enters the domestic sphere:

> The computer must find its place within the household's routines of activity, and within the structures of interpersonal relationship between household members, it must also establish a relationship to both the physical layout of the house and other household objects. (Lally, 2002: 17)

As Lally discovers, 'patterns of computer use [. . .] are constructed in complex ways' (2002: 173).[15] Between household members there are ongoing and shifting negotiations over ownership (which Lally conceives in process-based rather than contractual terms), location, as well as control of these technologies. Commenting on the 'patterning of computer use', Lally notes:

> Over an extended period of time [. . .], temporal patterns emerge out of the ebb and flow of everyday household activity. These are also, however, actively managed to a significant degree as part of the parental role, and in particular illustrate strategies through which domestic structures are kept under control. (2002: 17)

Silverstone, Hirsch, and Morley (1994) refer to this as the 'moral economy of the household'. It is an economic system which 'involves the appropriation of [. . .] commodities [such as the home PC] into domestic culture—they are domesticated—and through that appropriation are incorporated and redefined in different terms, in accordance with the household's own values and interests' (1994: 16).[16] They also note that, as a result, 'it is the computer which is, as often as not, transformed by this incorporation, much more than the routines of the household' (20; see also, Bakardjieva and Smith, 2001).

This raises the issue (noted above) of how a teletechnology, such as a new computer, once incorporated into the domestic sphere, must 'establish a relationship to both the physical layout of the house and other household objects' within that space. According to Lally (2002: 3), 'although new technologies displace earlier ones, that displacement is neither complete nor simple'. There are rhythms that affect the incorporation into and ongoing uses of new technologies within the domestic home, and these can and usually do change considerably over time. Thus, Lally argues, it is important to understand the domestic environment in the following relational terms:

> The computer [to stick with this particular example] cannot [. . .] be seen simply as an independent object within the home, but is one which forms one node in a network of objects which are disposed within the household in relation to each other. It is just one element in the total ensemble of domestic objects. (Lally, 2002: 175)

This in turn has significant implications for how we understand the domestic home as a site and home-making as a practice:

> Within the physical space of the dwelling, the material home is constructed through the assembly and configuration of objects, such as furniture, decorative items and technologies. These organizations are not static, but interact dynamically with those who inhabit them, as the material substrate to their patterns of everyday life. (2002: 10)

These last two statements by Lally hold twofold significance. First, they develop a picture of the domestic 'home' as a complex site of internal and ongoing negotiations. Secondly, they strengthen the arguments made by Allon, McQuire, and Tabor regarding the interpenetration of the domestic home by ICTs. They do this by positioning the domestic home as a 'permeable' site formed through a complex system of internal negotiations *vis-à-vis* equally complex external forces and wider practices of consumption.[17] In a word, home is not—or is not strictly—the tightly enclosed space of privacy and exclusion it is often depicted to be (if indeed it ever really was). This finding connects with Doreen Massey's important formulation of place as relational. As Massey writes, 'a large component of the identity of that place called home derived precisely from the fact that it had always in one way or another been open; constructed out of movement, communication, social relations which always stretched beyond it' (1992a: 14). This particular conception of 'home' is valuable for understanding how home is affected by information and communication technologies.

The full extent of the interpenetration of public and private via electronic media—and of the home as a complex site constructed out of the 'movement, communication and social relations' which stretch beyond it—is evident in and captured by Silverstone and Haddon's 'domestication' model of technology consumption and use. This is summarised below.

Mobile Privatization and the Domestication of Technology

Silverstone and Haddon develop the 'domestication' approach as a way of making sense of the 'intimate relations' that characterise the 'production and consumption of a new media and information technology' (1996: 54). Their interest in the domestic emerges from a belief that it is difficult to think of domesticity without making reference to the increasing presence of media and information-communication technologies in the domestic home. But, they add, the reverse is also true: 'No account of technological innovation can ignore the particularity of that domesticity and the processes by which it is sustained' (Silverstone and Haddon, 1996: 61). Thus, as Silverstone and Haddon understand it, domestication takes on a double sense. It refers to the home as a socio-technical site for the consumption of new (and old) technologies, as well as constituting a particular method, or model, for making sense of the processes by which these new technologies are consumed and 'domesticated' or naturalised within and beyond this site. According to the second of these two understandings of this notion, they argue that domestication becomes a 'more or less continuous process in which technologies and services are consumed [. . .] and, through the process of consumption, are given meaning and significance' (67).

This functions according to a double process, in which the domestication of new technologies involves a 'taming of the wild and a cultivation of the tame' (60). New technologies (such as computers, DVD players, and mobile phones) are considered exciting but also potentially threatening and in need of being 'brought [. . .] under control by and on behalf of domestic users' (60).[18] Yet, as soon as they are 'domesticated' through ownership and appropriation into the culture, flows, and routine of family, household, and everyday life, these technologies are cultivated (60).[19] That is to say, as they become familiar, or as they are placed alongside or replace existing technologies, the uses of these technologies change and are redefined (68).[20]

This understanding of technological domestication owes a debt to the earlier work of Raymond Williams, and especially his idea of 'mobile privatization' (Williams, 1992: 20). Williams developed this concept as a way of encapsulating a complex series of technological developments, which he saw as characterised by 'two apparently paradoxical yet deeply connected tendencies of modern urban industrial living: on the one hand mobility, on the other hand the more apparently self-sufficient family home' (20).

In their reworking of his concept, Silverstone and Haddon focus on the second of two tendencies that Williams describes, giving detailed consideration to the complex processes by which technological developments are integrated into and 'domesticated' or 'naturalised' within the domestic environment.

Nevertheless, the strength of the domestication model lies in its wider application. The authors' investigation of new technologies and how they are incorporated into the patterns of everyday life is important in that it extends the notion of domestication beyond the confines of the traditional domestic home. This has two benefits.

First, it intersects with, or allows parallels to be drawn between, wider geopolitical considerations of mobility and the ongoing importance of home. This is evident, as discussed earlier in this chapter, in Derrida's remarks on the 'powerful [. . .] recourse to the at-home, the return toward home' (Derrida and Stiegler, 2002: 80), and in Scott McQuire's notion of the 'technological uncanny' as it connects with what he calls 'the displacement of geopolitics' (1997a: 700). These formulations, as well as more general understandings of the 'home' and the 'domestic', are both significant in trying to understand the complexities of home and teletechnologies, and the interactions between networked mobility and place.

The second benefit of a more expansive understanding of the processes of technological innovation and consumption is that domestication becomes an elastic concept with wide application for understanding various forms of technological innovation and use, including how we might understand networked mobility and its uses.

For instance, in advancing how new technologies are domesticated, Silverstone and Haddon argue that the functions of certain technologies may, when incorporated in the home or household, be 'somewhat different from those intended by designers or advertisers' (1996: 64). They may also change over time (64). In addition, they also note that households are 'conventionally and habitually quite adept at a kind of seamless shifting from one technological input and resource to another as well as being adept at their simultaneous use' (66). Both observations are supported by empirical research into mobile phone use. Unintended use can be observed in the practices of 'immobiling' observed by Yoon (to be discussed below). Shifting between and simultaneous use of various technological resources is evident in the widespread practice of incorporating both household of fixed or landline and mobile phone connections in the routines of everyday life.

It is this wider context of the practice of everyday life that also has led to the expansion of the domestic home as a site. Silverstone and Haddon make this point in passing near the end of their study:

> Household boundaries are extended [. . .] by the increasing mobilization and personalization of communication and information technologies, as walkmen and mobile phones offer a new kind of nomadic access and media participation, constant availability and increasing dispersal of information consumption. (Silverstone and Haddon, 1996: 69–70; see also, Spigel, 2004)

This observation, in effect, forms the point of departure for David Morley's (2003) examination of the uses and impacts of networked mobility. Drawing on Silverstone and Haddon's model of domestication, Morley takes up an issue that is remarked on, but is otherwise left undeveloped by Silverstone and Haddon: that is, the contradictory dynamic or tension between processes of technological domestication which occur within the family home, and other practices of everyday life by which mobile phone use 'dislocates' (in the sense of a relocation rather than a supplanting of) domesticity.[21]

Morley's consideration of this 'dynamic' is of particular interest, given the focus on home in the present chapter, as well as the overarching concern of this book with questions of place.

As was discussed in Chapter 3, place is a much maligned notion within contemporary critical discourse. It is criticised for its lack of definitional precision; it is linked to strategies of exclusion; it is seen as marginal to modernist considerations of time and space; and with the emergence of cyberspace and virtual community, it is said to be left behind or reduced to the status of metaphor. Yet, place is a resilient notion and persists in the face of all these continuing challenges.

MOBILISING PLACE

Here I examine the notion of place in relation to networked mobility and mobile phone use in order to ask what relevance, if any, place has in the context of networked mobility. Does mobility render notions of place obsolete? Or does place persist? And if the latter, what happens to the common conception of place as a 'proper, stable, and distinct location' (Morse, 1990: 195) as a result of mobile practices? Do these technologies and practices contribute to altered understandings of place?

The notes on place and globalisation offered in Chapter 3 provide a valuable initial context for the ensuing discussion of place and mobile technologies. This is for two reasons. First, because questions of mobility and place (especially local place), like questions of CMC and place, are situated firmly within globalisation debates and address shared concerns. As Larissa Hjorth (2005: 208) writes, 'The dynamic interaction between globalisation and practices of locality is nowhere more apparent than in debates surrounding mobile telephony and its dissemination and appropriation at the level of the local'. Secondly, this context provides a good example of how place, while imprecise in definition, is complex in understanding and application. This is also the case with networked mobility. Place is an important notion in studies of networked mobility, albeit one which is employed in rather general terms as a somewhat vague expression of geographical experience (to paraphrase Edward Relph).[22] Moreover, how place is understood through and shaped by networked mobility can be seen to be both complex and at times contradictory, and it is worth looking closer at place and networked mobility to tease out some of this complexity and contradiction.

Networked Mobility and Place

At one level, mobility in general and networked mobility in particular (bearing in mind that it is increasingly difficult to differentiate the two), both appear to contribute to a dislocation of place—or what Morley (2003: 439) refers to as the alleged 'death' of geography. Mobility in every form unsettles that which is considered to be fundamental to conventional understandings of place: its very stability. As Christian Norberg-Schulz (1980: 180) writes, 'human identity presupposes the identity of place, and that *stabilitas loci* therefore is a basic human need'. Arguably, *stabilitas loci* or the stability of place is even more directly unsettled by networked mobility: 'The mobile phone is often understood (and promoted) as a device for connecting us to those who are far away, thus overcoming distance—and perhaps geography itself' (Morley, 2003: 452). In other words, mobile phones are said to operate 'independent of place' (Wellman, 2001: 19). And where they are not exactly independent of place they appear immune to place, serving to insulate their 'users from the geographical place that they are actually in' by creating, as Morley puts it, a kind of 'psychic cocoon' around each user, much like a Walkman or an iPod does (Morley, 2003: 451 & 9). In light of these developments, it has been argued:

> [That] the importance of place as a communication site will diminish even more, and the person—not the place, household or group—will become even more of an autonomous communication node. (Wellman, 2001: 19)

The result, it is claimed, is 'the rise of networked individualism' (Wellman, 2001): the shift from 'place-to-place' communication to 'person-to-person' communication; or, from 'inter-household networks to interpersonal networks' (29–30).

But networked mobility enjoys a far more ambiguous relationship with place than is perhaps suggested by the above formulation. This is for several reasons. To begin with, 'despite all the talk of "postmodern nomadology" [. . .] most people's actual experience of geographical mobility' (Morley, 2003: 437) is still very limited. That is to say, 'global cultural forms still have to be made sense of within the context of what, for many people, are still very local forms of life' (437).[23]

For example, one curiosity about mobile phone use is the 'domesticity' that characterises much of the conversation that takes place via these devices. As Morley (2003) notes, 'What the mobile phone does is to fill the space of the public sphere with the chatter of the hearth, allowing us to take our homes with us, just as a tortoise stays in its shell wherever it travels' (452–453).[24] This is a point that will be returned to later.

There are also counter-intuitive uses of the mobile phone, such as when this technology does not so much transcend distance as 'establish parallel communications networks in the same space' (451). For example, studies of mobile phone use regularly report that, while it is not always considered acceptable practice, these devices are commonly used in same space settings, such as a school classroom or train carriage.[25]

Also contrary to the claim that networked mobility overcomes geography, is the prevalence of the question, 'Where are you?' (Morley, 2003: 440; Laurier, 2001), by which many mobile phone conversations begin (and many on-line chat room conservations also).

It is in this sense that mobile telephony responds to the writer Georges Perec's (1999: 83) lament that 'we always need to know what time it is [. . .] but we never ask ourselves where we are'. Perec's point, of course, is that even when we provide an answer—'we are at home, at our office, in the Métro, in the street' (83)—we only really 'think we know', and the answer betrays how very little we in fact do know about place, because the 'where' of 'some*where*' is tied up with the seemingly inscrutable workings of the everyday. However, a key argument of this chapter is that networked mobility prompts renewed consideration of the 'where' of everyday places by forcing us to reflect on our apprehension and comprehension of them in transit. More than this, networked mobility leads to transformed understandings—understandings which lend further weight to Doreen Massey's notion of a 'global sense of place' which posits that 'what gives a place its specificity is not some long internalised history but the fact that it is constructed out of a particular constellation of social relations [and, I would add, technological interventions], meeting and weaving together at a particular locus' (Massey, 1991: 28).

These transformations are evident in empirical studies of mobile phone use, which provide a clear indication of how place is experienced through and transformed by networked mobility.

In a study of Norwegian mobile phone use, for example, Ling and Haddon (2003) point to the key role the mobile phone plays in the 'micro-coordination' of everyday activities and, in particular, of basic daily travel arrangements.[26] 'The development of mobile telephony,' Ling and Haddon write, '"softens time" in that one does not necessarily need to agree upon an absolute point in time but rather can, to some degree negotiate, or micro-coordinate, over where and when to meet' (2003: 246).

The 'softening of time' through 'micro-coordination' is also strongly evident in studies of Japanese youth and mobile phone use (Ito, 2003, 2005; Ito and Okabe, 2005). These studies reveal that networked mobility has transformed the way that meetings are arranged in urban space. 'In the past, landmarks and times were the points that coordinated action and convergence in urban space. People would decide on a particular place and time to meet, and converge at that time and place' (Ito and Okabe, 2005: 226). Now, however, it is more likely that an initial and rather loose arrangement is agreed upon, and 'as the meeting time nears, contact via messaging and voice becomes more concentrated, eventually culminating in face-to-face contact' (227). It is also common for mobile communication to continue even after physical co-presence has been achieved in the same urban space (Ito, 2003). This elaborate series of micro-coordinations reveals a complex set of interactions and negotiations between place, physical co-presence and 'virtual' presence (see also Licoppe, 2004). One result, it is suggested, is that 'distant others are always socially co-present, and place—where you locate yourself—has become a hybrid relation between physical and wirelessly co-present context' (Ito, 2003; Wei and Lo, 2006). This would appear to complicate the idea of a shift from place-to-place to person-to-person communication.

Complementing these findings is Yoon's (2003) study of mobile phone use by South Korean youth. This study reveals other counter-intuitive uses of mobile technology that serve to further reinforce rather than diminish the importance of place, such as the practice of 'immobiling'. Yoon develops this term to describe certain strategies by which young mobile phones users 'immobilize' (turn off) their mobile phones in response to perceived sensitivities between peers concerning place, time, etiquette, and content (2003: 334ff). Turning off the phone also constitutes an important way of diminishing parental control by preventing parents from making contact via text or voice message. In both cases, 'immobiling' serves as a key means by which to develop 'local sociality' (329) and, in turn, 'retraditionalize the global' (340).

These studies and their findings are interesting in that they appear to validate Boden and Molotch's (1994) claim that we are influenced by an ongoing 'compulsion of proximity', and that technologies of distance do nothing to obviate the need for regular co-presence through face-to-face encounter.[27] In fact, one suggestion is that 'those who make the most phone calls are also those who interact with the largest number of people face to face' (Lévy, 1998: 32).

It also emerges from these studies that place—especially local place—is still central to the practice and understanding of networked mobility. But how local place is experienced through networked mobility is quite unique. It is a heavily mediated engagement, where place is experienced via a complex filtering or imbrication of the actual with the virtual. This is a key point, and will be expanded later at length.

To further appreciate the continuing importance of place to networked mobility, it is also valuable to consider 'how "mobile" traditions incorporate new technologies as they develop' (Morley, 2003: 443). The 'domestication' model, as developed by Roger Silverstone and Leslie Haddon and particularly as extended by David Morley, offers a useful frame for understanding this process.

As noted earlier, Silverstone and Haddon develop the domestication model in order to account for the complex processes by which new technologies are integrated into and 'domesticated' within the domestic home. The key contribution of Morley's essay on home and mobility is that it takes Silverstone and Haddon's model and applies it to networked mobility and mobile phone use, making explicit the connection that Williams drew in the early 1970s between domestication and mobility. However, in developing this link, Morley inverts Williams's earlier formulation. Raymond Williams (1992: 20) posits the idea of 'mobile privatization'—an idea which arguably reaches its apotheosis with the relatively recent advent of home theatres and the digitalised 'smart house' such as those discussed earlier in this chapter. Morley, on the other hand, inverts this formulation by considering contemporary mobile phone use as a form of 'privatized mobility' (2003: 437ff). As such, networked mobility further extends household boundaries by 'dislocat[ing] the idea of home, enabling its user, in the words of the Orange advertising campaign in the UK, to "take your network with you, wherever you go"' (451). In so doing, networked mobility reinforces the idea of home in a domestic sense (and with it, the appeal to what Derrida terms the 'at-home'). These ties to the domestic are further reinforced by the provision included within many mobile network billing plans for the user to have reduced rates (or free calls) between a 'home' landline and a mobile phone.

Mobile Home: the 'Dislocation' of Domesticity

Complementing the domestication approach to 'historicising' the ongoing relevance of place and home to networked mobility is the equally illuminating context of 1960s 'experimental architecture'. This brings us back to the work of Peter Cook and Cedric Price, first introduced in Chapter 4. For example, in Peter Cook's (1970) study, *Experimental Architecture*, we find the same ingredients examined by Silverstone and Haddon and also by Morley: domesticity, technology, mobility, and place.

In Cook's survey, a key point of departure is, as he describes it, the 'turn to eclecticism' in the architecture of the 1960s:

> The word [eclecticism] had derogatory implications only a decade ago (architecture was to be pure and discriminating), but it now implies a

positive openness and absorption of anything that might be useful to a project. (Cook, 1970: 14)

This 'bricoleur' approach to design is most strongly felt in the rapid uptake at Cook's time of new buildings materials (due to advances in materials manufacturing technologies) and engagement with and 'absorption' of telecommunications and media technologies (due in large part to the influence of the writings of Marshall McLuhan). The 'opportunity of the material' (Cook, 1970: 55) and the turn 'towards technology as a great force for a new architecture' (125) dovetail in two interconnected concerns that are central to Cook's study: the first is the potential of telecommunications technologies to transform the domestic house; the second is an abiding interest in mobility (and neo-nomadism).

To address the first of these interconnected concerns, a key reason for such a strong renewal of interest in the function and operation of the house in 1960s experimental architecture, beyond developments in materials mass production, is found in Cook's realisation (after McLuhan) that 'communication is becoming as powerful as [a] tactile or representational environment' (1970: 125). And in a statement that arguably foretells the soon-to-be-reality of computer-mediated networked communications, Cook writes, 'we shall reach a point quite soon where real time and imagined dimension can be made to interact' (126).

Commensurate with this realisation is an acknowledgment of the effect of communication on the fabric of the traditional family unit (128). As Cedric Price (1984: 48) writes, 'the house is no longer acceptable as a pre-set ordering mechanism for family life'. Price's concern is in questioning the taken-for-granted *function* of the house in light of the aforementioned developments in mass media technologies. The domestic house becomes in Price's terms 'a 24-hour economic living toy' (48)—a kind of miniature domestic 'fun palace'.[28] Price's conception of a technologised and functionally open house of experimentation did find some form of architectural expression some time later in the *House of the Century* (1973) project by the Ant Farm collective of US architects and media artists. This humorous experiment in future living—described as a 'ferro-cement domicile with futuro-phallic features' (Seid, 2004: 25)—was constructed beside Mojo Lake, Angleton, Texas, and featured an array of electronics devices and equipment for the media-savvy occupant. Price's domestic vision of the house as a '24-hour-economic living toy' and Ant Farm's attempt to realise such a vision both constitute precursors to the modern 'electronic house' and examples of what Scott McQuire (2003: 103) describes as the 'repositioning of the home as an interactive media centre'—key developments in the continuing 'domestication of technology'.

The second of the two interconnected realisations in Cook's study reflects the kinds of techno-social transformations described elsewhere in this book. That is, global networked telecommunications technologies, coupled with burgeoning global travel and interconnected financial markets, not only increase the sense of an increasingly shrinking planet (McLuhan's 'global village'), but also animate a shift from the traditional conception of a *stabilitas loci* towards a culture of mobility (*mobilitas loci*) (see Urry, 2000a, 2002). Or as Cook (1970: 131) puts it: 'the future environment will be where you (yourself) may find it' (see Figure 6.2). Implicit in Cook's understanding and in much of the experimental work gathered in his survey, with its emphasis on technology and mobility, is the belief that 'place' is antithetical to technology, and vice versa. Yet, as Cook readily admits, what is being transported in these experiments with technologies of mass fabrication and mobility is, precisely, the house:

Figure 6.2   This cartoon by Cedric Price captures a shift in attitude towards a culture of mobility and the idea that 'the future environment will be where you (yourself) may find it'.
Source: Cedric Price, architect, 1934–2003.
© Cedric Price fonds Collection Centre Canadien d'Architecture/Canadian Centre for Architecture, Montréal. Reproduced with permission.

There have been projects for the all-metal house, the all-plastic house, the all-paper house, the all-wooden house, the all-pneumatic house, the all-glass house, the house as a total dome, the house as a total box, the house as a total capsule. (Cook, 1970: 55; see also, Zuk and Clark, 1970)

Thus, what was motivated—albeit implicitly—by a desire to dislocate place from architecture and technology results, one might say, in the *dislocation of architecture in place/s*. As such, I would argue, this experimental architecture represents an important—if somewhat literal—precursory stage to the more recent transformations which Silverstone and Haddon (1996) describe as the 'extension' of household boundaries through 'increasing mobilization and personalization of communication and information technologies' (69), such as mobile phones, and which Morley (2003) theorises as the 'dislocation of domesticity'.

The further import of these experimental investigations into mobile structures lies in the fact that they can also be seen as part-and-parcel of a 'basic dialogue between movement, structure, and the possible transfer of events and their location within the structure' (Cook, 1970: 101). 'The sift of these three conversations,' Cook writes, 'can be passed across most experimental projects' (101). Simply put, this interest in mobility is not just for mobility's sake. As Cook says of Archigram's (1968) *Ideas Circus* project, but which might be taken as a more general summary of the motivations behind these architectural experiments in mobility, there is an underlying consideration of 'questions of place, facility, equipment and the idiosyncrasies of the users' running throughout most of these projects (Cook, 1970: 122).

From *Stabilitas Loci* to *Mobilitas Loci*: Networked Mobility and the Renegotiation of Place

And so it is with networked mobility. These very issues—of place, facility, equipment, and the idiosyncrasies of use—are also at stake in networked mobility's engagement with and renegotiation of place. This reiterates what is a key point: place persists but does not remain unchanged by these developments. That is to say, networked mobility in general and mobile phone use in particular, lead to altered or transformed understandings of place and place-making. These warrant further consideration here.

To begin to tease out how understandings of place are transformed by networked mobility, it is valuable to return momentarily to Ant Farm. In addition to their many dalliances with portable inflatable structures (Maniaque, 2004), Ant Farm also shared a deep interest in media technologies and associated issues of media representation. They explore this interest in a number of projects (including the aforementioned *House of the Century*), but especially in their *Media Van* (1971): an electronically equipped customised 1971 Chevrolet van in which the artists toured the United States. As Ant Farm recorded at the time, the impetus for *Media Van* was to 'realize our own brand of nomadism'. The idea of *Media Van* was one of 'total documentation', whereby, while trucking on the road, 'you have the capability to record via videotape, photographs, film, and mental notes what's going on around you' (Lord, Michels, and Schreier, 2004: 100). In short, *Media Van* utilised networked telecommunications technologies (including TV, audio, etc.) to document as thoroughly as possible the shifting and fleeting ambiences of place as they were experienced through mobility (see Figure 6.3). Such a deliberate artistic exercise in documentation of space and place has, more recently, rapidly become largely routine practice for many users of networked telecommunications technologies, especially those with picture phones. Indeed, so commonplace has this process of documentation become that it has given rise to the phenomenon of 'life-caching': the use of digital cameras and picture phones and software such as Lifeblog to create digital diaries, scrapbooks, and photo albums which serve as mnemonic devices for sifting through and recollecting daily experiences (Schofield, 2004).

The extent to which place is transformed by mobility can be further understood by considering Marc Augé's account of 'non-places'. According to Augé (1995), the contemporary cultural landscape of globalisation is characterised by an overabundance of information and a growing tangle of interdependencies which leads to the creation of an 'excess of space correlative with the shrinking of the planet' (31). Augé coins the term 'non-places' to describe this expanding excess. 'Non-places' are those interstitial zones where we spend an ever-increasing proportion of our lives: in supermarkets, airports, hotels, cars, on motorways, and in front of ATMs, TVs, and computers. For Augé, such 'non-places' are the real measure of our time. The true extent of them can be quantified, he writes:

> by totalling all the air, rail and motorway routes, the mobile cabins called 'means of transport' (aircraft, trains and road vehicles), the airports and railway stations, hotel chains, leisure parks, large retail outlets, and finally the complex skein of cable and wireless networks that mobilize extraterrestrial space for the purposes of a communication so peculiar that it often puts the individual in contact only with another image of himself. (Augé, 1995: 79)

In short, Augé argues that it is not mobile privatisation per se that is new, 'but the interpenetration of layer upon layer of built environment and representation, the formative and derivative, the imaginary and mundane' (Morse, 1990: 210). It will be recalled from the previous chapter that this same process has been described elsewhere as the overlaying of a 'third nature' of information flows on the 'second nature' of cities, harbours,

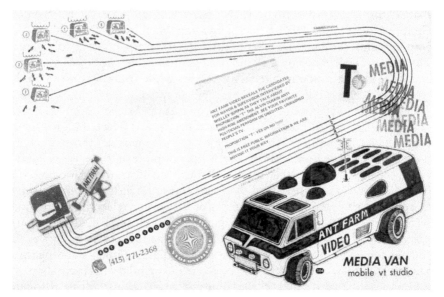

Figure 6.3 Ant Farm's *Media Van* project (1971) used a Chevrolet van and net-worked communications technologies to document as thoroughly as possible the shifting and fleeting ambiences of place as experienced through mobility. Such a process has largely become routine practice for users of contemporary networked telecommunications technologies, especially picture phones.
Source: Ant Farm, *Media Van (mobile vt studio)*, 1971. Ink, collage, rubber stamp, and sticker on card. 11 x 17 inches (27,94 x 43,18 cm). University of California, Berkeley Art Museum and Pacific Film Archive. Purchase made possible through a bequest of Thérèse Bonney, by exchange, a partial gift of Chip Lord and Curtis Schreier, and gifts from an anonymous donor and Harrison Fraker.

industry, and so forth, creating an 'information landscape which almost entirely covers the old territories' (Wark, 1994: 120).

In response to this seemingly overwhelming 'spatial excess', Augé makes two modest yet instructive and interconnected suggestions. The first is that such spatial relearning involves thinking about 'space as frequentation of *places* rather than a place' (1995: 85). As Augé argues, 'It is no longer pos-sible for a social analysis to dispense with individuals, nor for an analysis of individuals to ignore the spaces through which they are in transit' (120).

The second impetus is that we have to 'relearn how to think about space' (36). This suggestion forms a core point for development in the following chapter. For the moment, however, it is worth noting that implicit in the cata-logue of 'non-places' that Augé furnishes is the suggestion that this process of relearning requires an understanding of space as thoroughly technologised— or at least this understanding needs to be made explicit. Moreover, every-day engagement with these spaces, as the example of life-caching illustrates, involves the 'copresence of multiple worlds in different modes': the screen and the geography over which individuals travel (Morse, 1990: 206). The 'copresence of multiple worlds', Morse argues (1990: 203), presents a paradox for mobile experiences of place. On the one hand, these multiple mediations are characterised by a kind of 'detached involvement', a 'dreamlike displacement or separation from [one's] surroundings' (197). On the other hand, this multiple-mode of engagement might well hold wider promise. Morse writes, the 'task of reintegrating a social world of separated, dislocated realms is accomplished by means of an internal dualism, of *passage* amid the *segmentation* of glass, screens, and thresholds' (200).

Both sets of observations and considerations point to a significant shift in an understanding of (local) place. To recast Norberg-Schulz's formulation, it would seem a shift is being initiated from the notion of *stabilitas loci* or 'stable place' to what I have been terming *mobilitas loci*: the difference between place experienced as stable (if not fixed), to multiple places experienced in and through mobility. This shift, I would suggest, fills out Morley's (2003) understanding of the 'dislocation of domesticity'. It also connects with Doreen Massey's (1991) formulation of a 'global sense of place'—but more on this later.

To conceive of place in this way is to come almost full circle in an understanding of how place is experienced: from the 'mobile gaze' of the nineteenth-century, via what Anne Friedberg (1993) terms the 'virtual mobile gaze' of late twentieth century postmodernism, to what might be understood as a 're-mobilised (virtual) gaze' with the advent of mobile (particularly image-enabled) telephonic technologies.

Networked mobility does prompt a renegotiation of place, much as strolling (*flânerie*) and the 'technologised' spaces of the grand arcades did in the nineteenth century. With networked mobility, place-based renegotiation takes a number of forms: from individual (usually pedestrian) navigation of (largely localised) place/s, to broader perceptual considerations concerning the navigation of place via a re-mobilised, distracted (virtual) gaze, and the documentation of place through mobile phone cameras and the related practice of 'life-caching'. Thus, rather than 'liberate' us from place, as Wellman would have us believe, these technologies arguably refocus the individual on the fluctuating and fleeting experiences of place/s and their impact on the fabric of everyday life.

Mobilitas loci—the renegotiation of place via networked mobility, and the interrogation of 'questions of place, facility, equipment and the idiosyncrasies of the users' that this renegotiation prompts—generates manifold questions concerning the apprehension and examination of place through networked mobility. For example, the increasingly mediated nature of our engagement with place—especially via mobile telephony— would seem to suggest the need for some kind of hybrid approach to visual perception (at very least) which bridges established understandings of landscape structure and perception (Higuchi, 1983), with more recent analyses drawn from virtual reality, cinema studies, interface design, phenomenology, and other sources (see Richardson, 2005; Cooley, 2004), as well as incorporating (or accounting for) other sensory data (Barker, 2006; Bull, 2001, 2004). Morse's notion of the 'copresence of multiple worlds in different modes' experienced as an 'ontology of everyday distraction' is a productive step in this direction.[29] Indeed, Morse's work highlights the very impossibility of maintaining an uncomplicated distinction between place in a strict or 'pure' geographical sense and mediated experience (and construction) of it.

Furthermore, *mobilitas loci* and practices of networked mobility raise crucial questions about the fate of the social in an increasingly networked, mobile world—questions that impact on the arguments of this book, and which are to be taken up and addressed in the final chapter to follow.

MODULATIONS OF ENGAGEMENT AND PLACE-BASED TRANSFORMATIONS

One of the key arguments concerning the impact of information and communication technologies on the domestic home has been that these technologies significantly erode the distinction between inside and outside, and especially the distinction between the public and the private.

What emerges from these examinations is an understanding of the domestic home as a key yet complex and contested site—one that is important for understanding the detailed and contradictory processes by which new technologies are 'naturalised' or 'domesticated'; one that is constructed through a complex interplay of public-private negotiations, outside forces, and private influences and uses; and one that remains crucial for understanding the contemporary interactions of and interconnections between place, 'community', and teletechnology.

Studies of the incorporation and consumption of domestic technologies and electronic media within the home—their 'domestication'—suggest the role of these technologies is ambiguous. On the one hand, they help in the negotiation of the various tensions which structure the domestic home—tensions which include those between public/private and inside/outside, as well as between comfort/discomfort, freedom/oppression, belonging/alienation, production/consumption, and so on. Meanwhile, on the other hand, these technologies simultaneously contribute to their very production and maintenance.

Prompted by the above, one initial proposition of this chapter is that it is perhaps more productive (not to say accurate) to approach contemporary teletechnologies—such as the home computer—as leading to a *heightening* or *intensification* of existing contradictory processes, rather than as a new phenomenon. To make this suggestion is by no means to understate or deny the real and significant impacts that contemporary teletechnologies are having on how we make sense of the domestic home as a site and the meanings associated with it. Indeed, to argue that teletechnologies intensify and further blur the tensions that structure the domestic home is a significant point, and one that carries important implications.

For instance, in a passage quoted earlier and worth repeating here, Fiona Allon suggests one way to account for the impact of the 'traffic of information flows' on the tensions between 'public and private, and their related dispositions, sociality and privacy, connectedness and seclusion' is to see these tensions as referring 'less to enclosed and defined spaces [such as the domestic home], than to *modulations* of engagement (frequently with streams of information both local and global) and to infinitely variable conditions of interaction and control' (Allon, 2001: 17–18—original emphasis). This is a productive approach in that it emphasises the 'porousness' of the domestic home as a site (it also has application for making sense of practises of networked mobility, as well as CMC and Internet usage). More crucially, it provides a way of framing things which emphasises processes of socio-technological engagement within and beyond this site, rather than the site and its localised setting alone.

This leads to the following points. First, that any understanding of the domestic home as an 'embryonic' community or 'virtual' (or independent) community needs to be measured against, and rethought in the light of, this alternative approach with its emphasis not on the site *per se* but as various 'modulations of engagement' within and beyond sites. To do so is to open the concept of the nuclear family as a form of 'community' to a critique similar to that made of other forms of 'community', such as the 'imagined community' of the nation-state, the 'local community', as well the very notion of 'community' itself (Massey, 2005: 180).

Secondly, Allon's notion of 'modulations of engagement' resonates with Doreen Massey's formulation of place as relational—what she calls a 'global sense of place'. As Massey writes, quoted here at length:

> what gives a place its specificity is not some long internalised history but the fact that it is constructed out of a particular constellation of social relations, meeting and weaving together at a particular locus. If one moves in from the satellite towards the globe, holding all those networks of social relations and movements and communications in one's head, then each 'place' can be seen as a particular, unique, point of their intersection. It is, indeed, a *meeting* place. Instead, then, of thinking of places as areas with boundaries around, they can be imagined as articulated moments in networks of social relations and understandings, but where a large proportion of those relations, experiences and understandings are constructed on a far larger scale than what we happen to define for that moment as the place itself, whether that be [a home,] a street, or a region or even a continent. And this in turn allows a sense of place which is extroverted, which includes a consciousness of its links with the wider world, which integrates in a positive way the global and the local. (Massey, 1991: 28)

This important, and in many senses revolutionary, understanding of place (Cresswell, 2004: 53–79) has implications for many of the key concluding arguments of this book. In the present context, though, the passage is significant in that it predates much of the key contemporary research on mobile technologies by several years yet successfully captures the main argument developed earlier in relation to networked mobility and place. The core aspects of this argument are reiterated below.

For a number of critics writing on networked mobile technologies, these technologies are claimed to operate independent from and immune to place, contributing to what David Morley refers to as the alleged 'death' of geography.[30] Geographical displacement is said to be the result of a shift, inaugurated by these mobile information and communication technologies, from 'place-to-place' connectivity to 'person-to-person' connectivity. In other words, mobile phones—and related forms of portable, networked technologies—are said to privilege the individual over place, and over the communal (Tofts, 2005).

The response to these claims developed here has been drawn from two separate sources: existing empirical studies of mobile phone use, and the theoretical framework known as the 'domestication' approach to understanding the consumption of domestic technologies. Studies of mobile phone research contradict the above assertions regarding geographical displacement insofar as 'place' is seen to have ongoing relevance to this research and, indeed, plays a crucial part in how mobile technologies are understood to be used on a day-to-day basis.

The 'domestication' model of technology consumption offers a number of important insights that bear repeating here. For instance, in developing this model, Silverstone and Haddon, and Morley, argue that the domestic home remains a crucial site in the consumption of information and communication technologies, including, as Morley makes clear, in the consumption and use of mobile phones. (And thus, the domestic home also remains an important site for understanding broader issues of human-technology interaction, as does the more general trope of 'home'.)

Morley also counters the prevailing view that mobility leads to the 'death' of geography and the decreased relevance of place. While he argues that mobility 'dislocates domesticity', he means this in a very precise sense: a shift from a single location (home), to multiple locations; of 'home' as something that is 'transported' while simultaneously and continually orienting these practises of (localised) mobility. That is to say, the 'place' of 'home' (which can be taken here in its physical and its broader psychological sense) 'haunts' networked mobility.

What should be stressed, though, is that in addition to the continuing relevance of the domestic in understanding networked mobility, and the claim that place remains relevant to mobile phone use, place is also actively transformed by networked mobility and mobile phone use. These transformations represent a shift from the traditional conception of *stabilitas loci* (stable place) to *mobilitas loci* (the renegotiation of place via networked mobility, and the interrogation of 'questions of place, facility, equipment and the idiosyncrasies of the users' that this renegotiation prompts). *Mobilitas loci* raises the issue of how place is perceived and experienced phenomenologically through networked mobility. In addition, there is the question of how best to theorise this process. In response to this second question, it was proposed that Morse's notion of the 'copresence of multiple worlds in different modes' experienced as an 'ontology of everyday distraction' might offer a productive initial model. Morse's (1990) model acknowledges the hybrid and heavily mediated nature of this engagement and, indeed, highlights the very impossibility of maintaining an uncomplicated distinction between place in a strict or 'pure' geographical sense and mediated experience (and construction) of it. As Bill Mitchell (2005: 182) puts it, teletechnologies such as the mobile phone 'embed the virtual in the physical, and weave it seamlessly into daily urban life'.

Such an understanding represents a significant departure from the way in which teletechnologies were understood by many critics in the 1980s and early-to-mid-1990s, where a dichotomy often prevailed that opposed 'virtual' space to 'physical' or 'real' space, and conceived of teletechnologies such as the Internet as either 'utopian' or 'dystopian'. Even with such a shift in attitude there is nevertheless still scope for, and a definite need to develop, (a) a broader, more flexible formulation of 'place' and, in conjuction with or as part of this, (b) a more thorough theoretical account of how these understandings and experiences of place intersect with and are informed by teletechnologies. These are issues that run throughout this book, and are brought together in the final chapter to follow.

7 Rethinking Teletechnologies, Place, and Community

'We will have to find a way of housing virtual space in real space. The virtual is the antithesis not of the real but of the actual.'

—Paul Virilio (2001: 146)

'Community identification is necessary to self-identification, and both are necessary to any politics of difference.'

—Honi Fern Haber (1994: 125)

'We are always, inevitably, making spaces and places.'

—Doreen Massey (2005: 175)

A wide and diverse discursive terrain has been traversed in this examination of the literature on teletechnologies, place, and community. This book initially proceeded on the basis that place and community have, historically at least, been considered important concepts for making sense of who, what, and where we are. It was also understood that the ongoing role and significance of these two concepts is less certain in the present age of global telecommunications networks and technologies of distance. In light of this context, examination sought to develop a better understanding of the interactions between teletechnologies, place, and community in the electronic age. It also sought to reveal and make sense of some of the discursive, rhetorical, and other textual strategies that characterise writing on place and community in the literature on teletechnologies and place-making.

From this examination, place emerges as a persistent but much-maligned notion. It is often recorded in writing as peripheral to how teletechnologies (such as the Internet) are conceptualised and experienced. For example, in writing on computer-mediated communication, the notion of place is reduced to the status of metaphor. These metaphors are vital to how teletechnologies such as CMC are imagined and described, as they are to how computer interfaces are designed. Thus, they are considered useful to the extent that they provide a temporary sense of 'spatiality' to on-line social interaction, but are in many cases ultimately considered extrinsic or supplementary to this interaction, and in some instances are actively resisted. For these and other reasons, then, one of the key discursive strategies in writing on teletechnologies (especially from the 1980s and 1990s) is to frame place as a limit to be overcome.

Meanwhile, with respect to the notion of community, it, too, emerges as a persistent but much maligned notion. For example, in the early 1990s, it gained currency in the literature on teletechnologies as initially an expedient metaphor for describing the social interactions made possible by computer-mediated communications technologies. Increasingly, however, community is considered a problematic and contested term, both within and outside of writing on teletechnologies. Nevertheless, there is widespread recognition of the need for some kind of being in common or 'community', but substantial disagreement over what form/s this might or should take.

In short, place and community are both complex notions (although this complexity is not always acknowledged, especially in everyday usage of these terms), and continue to have relevance for understanding the complex interactions with and social uses of teletechnologies.

By the same token, teletechnologies, including both how they are theorised and used, also impact on, and some cases actively shape and alter, understandings and experiences of place and community. Because of this interaction, both the concepts of place and that of community require significant further conceptual refinement or reframing. This is because the more conventional and widely accepted understandings of the idea of place and community are increasingly inadequate for accounting for the complex nature of teletechnology interaction and its impacts in the present electronic age. Alternative conceptions of these terms are required, both in order to draw out the internal complexities of these ideas, and in order that they more productively contribute to and enrich our understanding of teletechnology use and a wider techno-socio-spatial *milieux*.

In light of this close interaction and inter-effect between teletechnologies, place, and community, there is also a need for a fuller, more conceptually rigorous model for accounting for the interactions between teletechnologies and 'actuality'. This is required in order to move beyond the erroneous and limiting yet persistent popular opposition of the 'virtual' versus the 'real' in accounts of teletechnological engagement.

Any alternative model faces three challenges. First, it must account for the increasingly extensive interactions between different forms of community and layers of technological mediation, whilst remaining as open as possible to the radical alterity of others and recognising the need for some form of (non-restrictive) 'being-together-in-community'. Second, it must present new ways of thinking about place in non-restrictive terms, and which account for its heavily mediated nature, the way that place persists in, but is also transformed by, the whole apparatus of teletechnology and its uses. Third, it must be a model that is not only permissive of political, social, place-based, and technological change, but sees such change as a structural necessity and an ongoing imperative.

In responding to these challenges, a three-part proposal is developed below that draws (in the main) from the theoretical work of philosophers Jacques Derrida, Honi Fern Haber, Todd May, Jeff Malpas, and the work of geographer Doreen Massey. This proposal provides for alternative and philosophically coherent conceptions of place and community, and develops a more nuanced and flexible account of the interactions between teletechnologies and actuality, and their intersections with ideas of place and community.

(1) ACTUVIRTUALITY AND THE 'FUTURE-TO-COME'

The first part of the proposal responds to a genuine need for a more thorough, and nuanced, theoretical framework (or model) that accounts for the full reach and interpenetration of what has been described as the 'compelling tangle of modernity and technology' (Misa, 2003). This can be found (at least provisionally) in Derrida's notion of 'actuvirtuality'. Some contextual background to this notion is required here.

There has been a tendency in earlier writing on computer-mediated communications technologies and the Internet (especially during the 1980s and early-to-mid-1990s) to frame these technologies according to a strict binary (either/or) logic. As already reviewed in this research, in the popular press in particular, there has been a tendency to juxtapose 'cyberspace' with 'real' space, 'virtual community' with more established forms of community, and so on. All too often these accounts also carried qualitative judgements about the 'virtual' being untenably 'utopian' or, alternatively, bleakly 'dystopian' (Rössler, 2001). In the latter case, this was often because these technologies were said to encourage a 'retreat' into cyberspace and a corresponding disengagement from the 'real world' and its attendant complexities.

Ongoing, detailed research into contemporary teletechnologies, especially computer-mediated communications technologies, and their uses confounds such neat and tidy formulations. Quite simply, they do not hold up in the face of mounting research evidence. For instance, to draw from examples cited in earlier chapters, participation in 'virtual communities' overlaps in complex ways with other forms of social interaction, on-line identity formation is in many cases consonant with 'off-line' formations and personae, and, in broader terms, the very way in which we go about our daily lives involves complex negotiations between spaces and places and teletechnologies of all descriptions. Even to think of space and place as plural complicates these distinctions, as will be discussed in more detail in the third part of the proposal below.

In response, a number of theorists and commentators have sought to formulate more flexible frameworks which might capture these complex technosocial interactions, including Mitchell's (1999a) notion of an 'economy of presence', Castells's (1996) concept of a 'space of flows', Morse's (1990) discussion of an 'ontology of everyday distraction', and Simon Marvin's (1997) thinking on electronic and physical links in telecommunications and urban environments.

Yet, the unproblematic embrace of binaries in framing and discussing teletechnologies and their social applications can be unsettled in a number of even more fundamental ways, three of which are sketched below.

The first way is to register that to draw distinctions between the 'real' and the 'virtual', and all the subsidiary distinctions that might flow from this, is at a fundamental level to maintain a belief in some kind of mythical rupture affecting modes of communication that came about with the emergence of the information age. Such a belief, Niall Lucy suggests, is erroneous:

> It is important to stress that since there has never been a time when the circulation of information was confined to face-to-face or person-to-person contact, then there was never any time at which it could be said that technology arrived within culture to 'displace' some sort of original and authentic mode of communication. (Lucy, 2004: 142)

Lucy refers to this elsewhere as a belief in 'autogenesis' (2001: 36). By this he means 'that one popular understanding of technology is that it brings about change at the expense of continuity; that it brings in the new on the basis only of an act of forgetting—forgetting the past', regardless of how this past might be conceived (36). Lucy's point is that 'technology is inseparable from human history—that is how we have recorded it, passed it on, kept it going. That's were we need to begin' (40). There is, it should be noted, a growing body of work that takes up this very task of 'historicising' emergent teletechnologies, placing these technological developments in historical and philosophical context, as part of a longer history of socio-technological innovation, adaptation, change and invention (Milne, 2010; Gitelman, 2008; Chun and Keenan, 2006; Gitelman and Pingree, 2003; Tofts, Jonson, and Cavallaro, 2002; Bolter and Grusin, 1999; Tofts and McKeich, 1998; Marvin, 1988).

The second point is that to maintain neat and uncomplicated binary distinctions—such as contrasting the 'real' with the 'virtual'—ignores the poststructuralist insight that in any apparent oppositional pairing the suppressed term in fact permits the possibility of the other, dominant term, because each term *contains* the other, contains elements of this difference. Take the utopian/dystopian distinction, for example. Any vision of 'cyberspace' as utopian must allow for the fact that it is also, and at the same time, dystopian—even if this alternative remains *in potentia*, so to speak. Krishan Kumar writes:

> Utopia and anti-utopia [or dystopia] are antithetical yet interdependent. They are 'contrast concepts', getting their meaning and significance from their mutual differences. But the relationship is not symmetrical

> or equal. The anti-utopia is formed by utopia, and feeds parasitically
> on it. It depends for its survival on the persistence of utopia. (Kumar,
> 1987: 100)

In other words, there is always a danger when trying to maintain an
uncomplicated binary distinction—that of forgetting that each term in an
oppositional pairing is 'contaminated' by the other and in fact permits the
possibility of the other. In this situation, the important issue—and it is
one that echoes the discussion of metaphor use in Chapter 3—is to *think
through* binaries, in both senses of the word through: as a way of simultaneously working *with* and *against* binary oppositions, of 'undoing-preserving' opposites as Spivak puts it (1997: xliii). Part-and-parcel of this
process of 'undo-preserving' opposites is remembering, as Meaghan Morris
remarks, that 'binarization, too, has differing aspects':

> Not all binarized terms are *opposed* terms (the drawing of a distinction
> is not necessarily oppositional), and oppositions can be alternating, or
> 'only relative' [. . .], as well as rigidly divisive; no positioning of a term
> is final, and 'nonsymmetrical' reversals are always possible between
> terms [. . .]. (Morris, 1996: 393)

The third and final example of how to unsettle the unproblematic embrace
of binaries in the framing and discussion of teletechnologies requires a
restatement of the issue at hand. As Hans-Ulrich Obrist writes, 'If we look
at how the Internet is perceived in the press and in the mass-circulation
dailies, this is where the big misunderstanding lies, there is always this
antithesis between virtual and real' (Obrist and Virilio, n.d.). This misunderstanding is said to be due to a fundamental yet persistent philosophical
'error'. As Paul Virilio explains:

> Virtual is not the opposite of the real, it's the opposite of [the] actual.
> It's [a] classic tenet of philosophy, reality has two faces, one actual,
> proceeding to the act, and one virtual, that which is potential. Virtual
> is the opposite of actual, not real. (Obrist and Virilio, n.d.; Virilio,
> 2001: 146)

In addition to the other responses detailed above, it is this emphasis on
the relationship between the virtual and the actual which offers a particularly productive basis from which to begin to respond to and account for
the complex interactions between teletechnologies, place and community.

The philosophical nature of the relationship between the 'virtual' and
the 'actual' has been examined at length by a number of recent thinkers,
most notably by Gilles Deleuze (1994, 2002; Williams, 2003: 198ff) and
Pierre Lévy (1998; Grosz, 1998). The focus of Deleuze's work in this area
is an extended critical engagement with the philosophical understanding

of the actual/virtual dichotomy as it is classically understood; his treatment of the 'two faces' of reality is claimed by some to represent a 'treasure
trove for new media' theorists and practitioners (Miles, 2002; see Savat and
Poster, 2009). Lévy draws inspiration from and further extends the ideas of
Deleuze by examining this classical philosophical opposition in relation to
technologies of the virtual, such as cyberspace.

Notwithstanding the above, it is Jacques Derrida's consideration of the
whole problematic of actuality that is focused on and explored here. The
reason for this is fourfold. First, Derrida (like Deleuze) is concerned with a
philosophical investigation of both understandings of the virtual—the populist technological one, and the classical philosophical one—and how the
two understandings might be held in tension. Second, his examination of
the actual and the virtual directly involves teletechnologies as it is framed
within a philosophy of 'ethico-political responsibility'. That is to say, he
specifically addresses philosophical questions of the actual and the virtual
(of that which proceeds to the act and of that which remains in potential)
in conjunction with questions of teletechnology use and the broader ethical and political impacts of and questions raised by both. Third, focusing
on Derrida's response to these issues (as opposed to Deleuze's or Lévy's)
is consistent with the focus of earlier chapters where Derrida's thinking
on other key and related issues has been consulted (including community,
metaphor, and the 'at home'). Finally, the decision to concentrate on Derrida's work is because his critique of 'actuality' in particular holds important implications for thinking about the interactions of teletechnologies,
place and community, and their 'place' within a broader geopolitical and
global context.

According to Derrida, 'actuality' consists of 'two traits', which he terms
'artifactuality' and 'actuvirtuality' (Derrida and Stiegler, 2002: 3ff). The
coining of these two 'portmanteau nicknames' (3) is a continuation of Derrida's deep-seated interest in 'thinking through binaries' and for revealing
what Niall Lucy (2004: 1) describes as 'the "aporetic" moment' or 'blind
spots' that structure all oppositional logic'. In this particular instance,
these issues are explored by Derrida around a consideration of the notion
of actuality and 'our experience of the present [. . .] as something that is
produced (made, made up)' (Lucy, 2004: 2). 'Actuality', Derrida writes, 'is
not given but actively produced, sifted, invested, performatively interpreted
by numerous apparatuses which are *factitious* or *artificial*, hierarchizing
and selective [. . .]' (Derrida and Stiegler, 2002: 3). Derrida coins the term
'artifactuality' to describe these processes.

The second trait of 'actuality', is captured in Derrida's insistence 'on a
concept of *virtuality* (virtual image, virtual space, and so virtual event)
that can doubtless no longer be opposed, in perfect philosophical serenity,
to actual reality in the way that philosophers used to distinguish between
power and act, *dynamis* and *energeia*,' and so forth (6). Derrida coins the
term 'actuvirtuality' to describe this second trait.

The importance of this insistence on the artifactuality/actuvirtuality of teletechnological experience is 'to let the future open' (21):

> It is to show that what counts as actuality in the present can no longer be confined to the ontological opposition of the actual and the virtual, despite the ongoing necessity of this opposition to every form of politics. (Lucy, 2004: 4)

Derrida's 'thinking through' of the traditional ontological opposition of the actual and the virtual (and of the actual as the 'undeconstructible opposite of artifice and the artefact' (Lucy, 2004: 4)) thus holds manifold implications for media and cultural theory that extend far beyond a simple commentary on the way in which media produce rather than simply record events. This reading necessitates a responsibility to analyse the media, 'to learn how the dailies, the weeklies, the television news programs are *made*, and *by whom*' (Derrida and Stiegler, 2002: 4). This responsibility is made all the more urgent for Derrida because, in our time, 'the least acceptable thing on television, on the radio, or in the newspapers today is for intellectuals to take their time or to waste other people's time there' (6–7).

For the concerns of this study, the value of Derrida's understanding of 'actuality' is threefold.

First, it explicitly acknowledges the way in which the media—or as Derrida terms it, the whole apparatus of teletechnology in general—is implicated in the construction of our experience of the actual. Crucially, Derrida's insights are not just limited to the modes of production and distribution of the media as traditionally understood. Their wider applicability can be grasped by considering both why and how he employs the term 'teletechnology'. While Derrida uses it in association with the media usually, '"the media"' is shorthand for what Lucy describes as 'several "tele-" effects and operations: "tele-communications, techno-tele-discursivity, techno-tele-iconicity"' (2004: 141). Thus, when Derrida writes of a responsibility and an urgency to analyse the media, this extends to all forms of media—including 'new media'. More crucially still, Derrida's insights into, and concern for, the construction of the actual extends to *all forms* of textual production, particularly those that attend and are shaped by 'tele' operations—including, for example, the discursive and rhetorical strategies by which a particular field (such as 'the virtual') is constructed, framed, imagined, discussed, and so on.

Secondly, Derrida's study is significant insofar as it suggests that teletechnologies work (alongside other factors) to unsettle any unproblematic distinction that may be maintained between virtuality and actual reality:

> Virtuality now reaches right into the structure of the eventual event and imprints itself there; it affects both the time and the space of images, discourses, and 'news' or 'information'—in fact everything

that connects us to actuality, to the unappeasable reality of its supposed present. In order to 'think their time', philosophers today need to attend to the implications and effects of this virtual time—both to the new technical uses to which it can be put, and to how they echo and recall far more ancient possibilities. (Derrida in Derrida, et al., 1994: 29–30)

Derrida's argument that teletechnologies work (alongside other factors) to disrupt any unproblematic distinction that may be maintained between virtuality and actual reality is particularly important in relation to architectural writing on cyberspace. As discussed in Chapter 5, cyberspace (or 'neuromantic') architectural writing displays a clear bias towards virtuality (both in its popular techno-spatial sense, and in its stricter philosophical sense) on the one hand, and an almost total lack of engagement with 'actual reality' and attendant questions of place and community on the other hand. Even in earlier architectural engagement with computers these last questions are often obscured (see Chapter 4). Yet, as Derrida points out, the whole apparatus of teletechnology is implicated in our experience of the actual. Thus, a productive but more difficult process is to consider both as they exist in tension: the actual and the virtual, the act and the in potential, the various technologically mediated and geographically dispersed communications networks and the various networks of places and people.

Lucy (2004: 154) suggests that, 'In trying to get us to think about the virtuality of the actual [. . .] Derrida is not suggesting that today's reality is some kind of postmodern illusion of the real, an effect of everyone spending too much time on the Internet and watching too much television'. On the contrary: 'We must bear in mind not only that any coherent deconstruction is about singularity, about events, and about what is ultimately irreducible in them, but also that "news" or "information" is a contradictory and heterogenous process' (Derrida in Derrida et al., 1994: 29).

While Derrida is certainly working here 'against a metaphysics of tidy distinctions' (Lucy, 2004: 154) between the actual and the virtual, the issue is fundamentally one of committing to and practising an 'ethico-political responsibility'. For example, in commenting on the role of the public intellectual, Derrida makes the following remarks that have resonance for all forms of intellectual engagement:

Asking oneself questions, including ones about the questions that are imposed on us or taught to us as being the 'right' questions to ask, even questioning the *question-form* of critique, and not only questioning, but thinking through the commitment, the stake, through which a given question is engaged: perhaps this is a prior responsibility, and a precondition of commitment. On its own it is not enough of course; but it has never impeded or retarded commitment—quite the reverse. (Derrida in Derrida et al., 1994: 40)

This call for responsibility of (and in) analysis has special relevance for a number of issues. It has relevance in relation to thinking about and questioning metaphor use. It has relevance for thinking about the construction of narratives, especially narratives about the history and future potential of teletechnologies. It also has relevance for thinking about how notions of place and community are understood and how they circulate in relation to writing on teletechnologies and place-making, not to mention how these terms are coopted and deployed in particular contexts (such as by urban planners, for instance, or within discourses of cyberspace architecture).

Furthermore, artifactuality and actuvirtuality not only necessitate a responsibility to analyse media, it is a responsibility that is open to the future and open to the other. As Lucy explains:

> Such an understanding of the actual as what is always 'actively produced' and 'performatively interpreted' is not an excuse for disengaging from public life or for affecting a disinterest in real-historical events. If the condition of actuality is that it must be made, then it must be able to be made differently [. . .]. That is why it's possible to make another artefact of the other—as the *arrivant*, the absolute stranger. (Lucy, 2004: 6)

In practical terms, Lucy suggests that 'remaining as open as possible to the radical alterity of others [. . .] means only that, as a place from which to start, an artefact of the other *arrivant* leaves open the greatest space for the possibility of a non-violent future to come' (2004: 6). This is an important point, yet it does however raise complex challenges for thinking 'community'—challenges to which we will return to shortly.

Thirdly, Derrida's critique of actuality is of value to the present study insofar as it offers hope and it encourages action—as one critic notes, 'Derrida's self-appraisal of his work was that it created the conditions to have dreams and to muster the strength to make a new world possible' (Turner, 2005: 135). His conception of the 'two traits' of actuality—'artifactuality' and 'actuvirtuality'—serves as a reminder or prompt 'to let the future open', to engage with what Gregory Ulmer (1994: 67) terms the 'possible impossible' of invention. As Lucy puts it:

> It is precisely because actuality is always produced that we have a responsibility to analyse its production and a responsibility to produce actuality ourselves—to make our own texts and artefacts out of what is going on in the world today. (Lucy, 2004: 154)

In other words, Derrida's critique of actuality is an encouragement to act, and to invent new actualities, especially those which remain as open as possible to the radical alterity of others. This, too, carries important implications for thinking about the future and compelling entanglements of technology, place, and community. It is significant in that it is productive and hopeful; it is productive because it is hopeful. It is also significant insofar as it is difficult, demanding as it does careful consideration of, and a sense of responsibility for, what this 'new' 'actuality' might be like and might involve.

Such a vision also has important implications for all forms of 'poetics'—meant here in the precise and inclusive sense in which Donald Theall (1995) conceives of 'the poetic' (xiii) as an organising term for all forms of 'inventive cultural productions' (7–8). Thus, the encouragement to act and invent—to act through invention—is inclusive of all forms of making or producing: architecture, urban planning, design and manufacturing, writing, and art—including as these use teletechnologies such as the computer.[1]

To make or produce in this context entails both difficulty and promise. On the one hand, the difficulty—which is the same as that which attends the utopian imagination in modern times—lies in 'identifying points of intervention in an increasingly complex social and economic structure, and of identifying the agents and bearers of social transformation' (Levitas, 1993: 258). On the other hand, the promise, as Theall (1995: 130) puts it, is that 'the poetic with its enigmatic revelation [. . .] invites discussion, assessment, and understanding rather than closing them off'.

These are just some of the implications and significance that the Derridean critique of actuality holds for this book.

To conclude this discussion of the uses and usefulness of this notion of actuvirtuality, it is important to add the following qualification. It needs to be stressed *what is not* being advocated in this idea of actuvirtuality for this project. This idea is *not* being promoted as some kind of totalising and abstract theoretical model. In other words, it is not promoted, as Roland Barthes (1986: 73) might say, 'in order to divinize it, to make it the deity of a new mystique'. Rather, the argument is that the Derridean conception of actuvirtuality furnishes a set of tools or prompts for thinking about, critically engaging with, and creatively responding to teletechnologies and our complex interactions with and applications of them.[2]

On its own, however, this is not enough. Two further facets or elements to this three-part proposal supplement and fill out this picture. The second part of the three-part proposal is to construct an adequate response to the problem of community discussed in Chapter 2. It is in response to these issues that discussion now turns.

(2) 'SUBJECTS-IN-COMMUNITY': TOWARDS A POLITICS OF DIFFERENCE

It has been suggested that 'the degree to which one can ground any kind of meaningful co-existence based upon a radical openness is questionable' (Cooper, 2004: 34). In this section, I will argue that it is possible to develop an understanding of difference that also goes a significant way towards offering a philosophically coherent and practical response to the doubt expressed above. This is the model of community developed by the philosopher, Honi Fern Haber. Before presenting her model of community, however, there is a need, again, for some contextual background.

As noted earlier in this book, the poststructuralist critique of the notion of 'community' argues that prevailing Western notions of community fail at 'remaining open to the other' (Lucy, 2001: 146). This points to the paradox—and thus the 'problem'—of community: how to formulate a model of 'community' that accommodates 'difference' (the 'Other'), while still acknowledging that 'Human existence is social existence [. . .]. Human being is a being-together-in-community'? (Secomb, 2003b: 9) How, in other words, to think afresh a model of sociality that welcomes the 'other' that threatens its very commonality?

These are difficult questions which have important and direct (as well as indirect) implications here. For instance, they impact on considerations of 'virtual community', on the positioning of the domestic home as an 'embryonic community' or 'independent community', as well as on the experimental projects of Cedric Price and others. Each of these examples is exposed to, and in some sense must therefore respond to, the poststructuralist critique of community.

To turn to the broader issue of teletechnologies and community, the extent to which our understanding of and participation in all forms of community are shaped by teletechnologies and technological mediation suggests the need for greater attention. This is not so much due to the semantic or conceptual differences between different definitions of community, and the problems facing each, but to singularities and particularities within a wider framework that is open to the interconnectedness of 'teletechnology-place-community' and which simultaneously responds to the aforementioned 'problem' of community.

There is certainly a growing meta-literature which addresses issues concerning methodological approaches to the study of CMC and community and which responds in part to this challenge. Terry Flew refers to this meta-literature as the 'new empirics' in Internet studies. 'I would propose', he writes, 'that we are now in a third stage of theorising virtual communities and CMC, which is very much an empirical phase' (Flew, 2001: 107). In simple terms, what this phase involves is a shift away from generalisations and broad theoretical pronouncements about 'the virtual' and 'virtual community', towards more detailed empirical work on the particularities of individual communities and their contexts. It also involves more detailed accounts of their interconnections with other community formations (Ward, 1998). This 'empirical turn' is most clearly apparent in books such as *Doing Internet Research* (Jones, 1999a), *Virtual Ethnography* (Hine, 2002), *The Internet: An Ethnographic Approach* (Miller and Slater, 2000), as well as in a number of individual papers (Slater, 2003; Lewis, 2001;

Burnett, 1999). What becomes clear from this shift in research emphasis, Flew suggests, 'is that the sheer diversity of forms of online discussion groups prevents any prior set of assumptions about the virtues of community, or of the "online" and "offline" worlds' (2001: 107).

This focus on singularities and individual contexts is an important and necessary development. However, no amount of detailed, micro-studies of this kind is likely to offer a satisfactory account of community and teletechnological mediation that simultaneously responds to the aforementioned 'problem' of community.

This is one of the key challenges posed by the poststructuralist critique of community for the information age: how to negotiate and formulate an understanding of the tensions that separate and bind singularities, places, and teletechnologies, in a way that is accommodating of difference yet still acknowledges the enduring importance of 'community', of 'being-together-in-common'?

Here the poststructuralist critique of community itself falls short. It seems unable to provide a satisfactory, practical response to the second part of the community paradox/problem: the issue of how to account for difference *and the ongoing need and necessity for a workable form of being-together-in-common, of 'community'*—let alone one which might encompass teletechnologies.

As philosophers Todd May (1997) and Honi Fern Haber (1994) argue, this failure of the poststructuralist response is due to, and hinges on, the following factors: its critique of structure, its rejection of unity, its exclusion of 'the possibility of coherent subjects', and its refusal of the possibility of community. According to Haber, the poststructuralist position holds that:

> Any structure—be it humanity, the subject, the social or cultural system, or history—comes into conflict with the law of difference because structure, traditionally understood, presumes to provide closure and coherence, unity or totality. Difference is repressed by structure. (Haber, 1994: 115)

This is in essence why community is resisted in poststructuralist thought, since community, it is argued, assumes consensus and unity ('communion', 'fusion-into-oneness'). Community and unity are thus seen as 'terroristic', a 'mark of terror' (38). That is, they lead to—or at least they have the latent potential for—'totalization' (4). And, as Haber goes on to note, 'totalization and totalitarianism are taken as synonymous' (116).

The notion of the 'subject' is intricately tied up in this critique. For, as Haber argues, 'the demand to universalize difference [. . .] demands that we foreclose on the possibility of the subject, be it individual or communal' (114).[3]

The unfortunate result, she argues, is that the poststructuralist maintenance of difference yields to a 'universalization of difference' (3). With this movement, the universalization of difference can become just as debilitating as that which it opposes. In Haber's words, it 'forecloses on the possibility of community and subjects necessary to oppositional resistance' (3).

A key shortcoming of this universalization of difference, Haber argues, is that poststructuralism fails ultimately in producing or enabling what she calls a 'viable' politics. On this particular point it is necessary to proceed with caution. To state this is not to suggest that poststructuralism is *apolitical*, which is a criticism that has long been levelled against it, on the contrary, in fact (see James, 2006: 152–173).

Rather, the point is quite a specific one: poststructuralism struggles to produce a 'viable politics' *insofar as* any such politics, as Haber explains, must 'allow room for consensus, community, or solidarity' (43): 'there is no politics without sameness and unity, even if that unity always has a remainder' (128). Or, as Todd May puts it, any viable conception of 'community' (which is part-and-parcel of a 'viable politics') has two requirements: 'that community be analysed in a way that avoids totalitarian conceptualizations and that it do so while still recognizing that individuality is in good part constituted by community' (May, 1997: 51). As May explains, this is the precise difficulty facing Nancy's philosophical conception of community:

> The conception he promotes cannot just hang out there as a way to conceive community nontotalitarianly [. . .], but, in order to have normative force, must be a conception that a community can consider as a reasonable possibility for itself to adopt. Unfortunately, no community can coherently embrace Nancy's conception of community as an antidote to totalitarianism, because in order to do so such a community would have to deny its own ratification of the value for which it would embrace the conception. (May, 1997: 41)

May is at pains to stress that he is not suggesting Nancy's conception of community is mistaken or self-contradictory, rather the problem lies with its possible realisation (1997: 43). In response to this failure, Haber in particular formulates what is understood here to be quite a productive model of what a viable 'politics of difference' (as she calls it) might look like.

In developing her politico-philosophical model, Haber does not reject poststructuralism—at least not completely. Rather, she draws key insights from it, which she then adapts:

> The metaphysical and ontological commitment to the law of difference (the inescapability of difference and the need for its recognition) generated by poststructuralist critique and presupposed in postmodern theory suggests [that] for a political theory to be viable, it must allow for the expression of difference. (Haber, 1994: 113–114)

Yet, in allowing for difference, there is a further caution: 'just as we must not universalize totality, so too must we avoid universalizing difference' (5).

'Each member of this team', she adds, 'will have to be reconceptualized to accommodate their synthesis' (5).

There are two key ingredients to this synthesis of solidarity and difference. These are, first, a belief in the importance of 'structure' (albeit according to a very particular understanding of this term), and, second, a commitment to the notions of the 'subject' and of 'community'.

In respect to the first of these ingredients, Haber points out that, 'To view structure as being hopelessly unjust or terroristic [. . .] is to hold on to the notion that there is something beyond difference or plurality' (117). Haber subscribes to the view that 'all structure is temporary and even artificial, and is always open to the possibility of being redescribed' (114). She also believes that 'difference can accommodate unity (structure) so long as unity recognises its subservience to difference' (117). But, further to this, Haber also maintains (in contradistinction to 'post-philosophies'):

> That any viable political theory necessitates structure—in particular it necessitates the need for the generation of subjects [. . .] and communities—even if these are themselves plural, internally inconsistent, open ended, and always amenable to deconstruction. (Haber, 1994: 114)

This in turn brings us to the second of her ingredients for a viable politics of difference—a reconsideration of the subject and community. Haber writes that a politics of difference is untenable unless it is able to 'allow room for consensus, community, or solidarity—though we need to understand these terms as something dynamic and essentially open to change, regroupings, and realliances' (43). This position is further elaborated below.

Haber acknowledges that 'radical pluralism is essential to oppositional politics' (114): 'oppositional politics must first of all demand that it be allowed to address concerns of the Other' (118). Indeed, as she notes, radical plurality is critical insofar as it 'makes alternative discourses an open possibility' (121). Yet, she is also quick to add that 'we must not allow the poststructural critique of language and the postmodern adoption of the law of difference to force us to conclude [. . .] that there is no subject' (120). Rather, the notion of the subject needs rethinking along the following lines: 'Poststructuralism can be read—or adapted to read—as necessitating only the claim that there is no autonomous, wholly self-creating, or coherent in the sense of single-minded or one-track self. The self can be many subjects' (120).

This formulation, Haber argues, has implications for how we think about community and, crucially, how we might think about community differently:

> Just as the self is a product of plural narratives, so too, 'community' is always subject to deconstruction. And the subject which is a product of one community will also be a product of many others, not all of whose

interests are compatible. However, we have vocabularies (though not only one vocabulary) only as a member of *some community or other*, and so *it is only as a member of some community or other than we are empowered.* (Haber, 1994: 120–121—original emphasis)

Or, to put this slightly differently, 'community identification is necessary to self-identification, and both are necessary to any politics of difference' (125). Very similar arguments to these are also developed by May (1997), where a central claim of his 'positive rearticulations' of difference and community is 'that a community is defined by the practices that constitute it' (52ff & 203–206).

Both of the above factors are captured in Haber's notion of 'subjects-in-community'. Simply put, this is the idea that 'there are no autonomous or non-plural subjects' and no singular and monolithic (that is, 'totalistic' and therefore 'terroristic') notion of community (Haber, 1994: 114). In combination, and as reformulated by Haber, these factors offer the basis for a politics of difference and for political change: 'Empowerment', she writes, 'comes about as the result of the struggle of subjects-in-community' (123).

Haber's critique develops a convincing response to both sides of the aforementioned community paradox/problem. On the one hand, it accommodates and is responsive to openness and difference. On the other hand, and at the same time, it (as does the closely related approach taken by May) takes the important stance that 'there is no politics without sameness and unity, even if that unity always has a remainder' (128). It achieves this balance in the following ways. First, it is 'non-restrictive'. That is to say, it acknowledges difference without universalising totality or, just as importantly, without universalising difference (5). Secondly, it is productive and hopeful (and politically engaged) in that it emphasises plurality and redescription ('iterability') *and* the inherent opposition of politics (2). And thirdly, it stresses that each of us is 'radically plural' (12), both as subjects *and* in our interaction and engagement with communi*ties*.

To conclude this discussion of community, it is important to outline the considerable value and implications of Haber's work, and her notion of 'subjects-in-community' in particular, for this project. The importance of her conception of a 'politics of difference' and of 'subjects-in-community' can be felt in different ways in different parts of this book.

To (re)conceive of community in the way that has been proposed above has manifold implications for how we think about this notion, both in general terms and in specific contexts, including poststructuralist theory, computer-mediated communication and 'virtual community', and the architectural and urban design literature on place-making.

In general terms, the model of community developed here is valuable in that it acknowledges the ongoing importance of some form of consensus and solidarity for human social existence. That is to say, to use Todd May's words, it 'avoids totalitarian conceptualizations [. . .] while still recognizing that individuality is in good part constituted by community' (1997: 51) and vice versa.

With respect to the poststructuralist position, this model builds on the insights and strengths of its critique of community, while simultaneously responding to the limitations of this approach (such as the threat or risk of a 'universalization of difference' and the refusal of the subject and of structure). For this reason, it is argued that a model which emphasises the plurality of subjects and community, as is advocated here, provides a more coherent response to the challenge of accommodating difference while still acknowledging that human existence is social existence.

It is important to stress that this particular approach to thinking community does not deny the ongoing risk or possibility of community in certain instances (re)turning towards 'fusion-into-oneness' or fragmentation. This risk will always be present in how the notion of community is understood (including in relation to place) and how it is sometimes coopted and utilised to specific ends. To think otherwise would be naïve, even dangerous. Rather, and this risk notwithstanding, the suggestion here is that this particular model of community offers hope and the grounds for a potentially viable 'politics of difference'.

For considerations of CMC and 'virtual community', this same understanding of the plurality of subjects and community (or 'subjects-in-community') is instructive insofar as it is 'democratic'. That is to say, it does not struggle to accommodate computer-mediated or 'virtual' communities in its ambit, nor does it require any differentiation between computer-mediated and other forms of community on structural grounds. This clears the way for an engagement with (and deconstruction of) community in all its forms that emerges from and in response to a philosophy of 'ethico-political responsibility' (see the first part of the proposal, above).

In overall terms, the biggest contribution made by the work of Haber and May is that it provides a coherent response to the aforementioned 'problem' of community. Thus, this reconceptualised account of community and the Derridean notion of actuvirtuality form key ingredients for developing a fuller response to, or account of, the interactions of teletechnologies, place, and community.

Yet, there is still one key element missing from this mix which complements the discussion of the first two parts to this proposal: the question of how to rethink *place* in productive and non-restrictive terms. This issue is the preoccupation of the third and final part of this three-part proposal below, where discussion focuses on the work of geographer Doreen Massey as well as that of philosopher Jeff Malpas.

(3) PLACE, RELATIONALITY, AND A
POLITICS OF CONNECTIVITY

The question of how place is understood and debated was discussed earlier in Chapter 3. From that discussion, at least four key issues emerged which contribute to place being considered a problematic and contested notion.

These are: (1) its definitional imprecision; (2) its strong associations with the local and the bounded; (3) the way in which it is often counterposed with space; and (4) the way it is often regarded as extrinsic or subsidiary to human experience and human subjectivity.

With respect to the issue of definitional imprecision, for example, Ed Relph observes that 'confusion about the meaning of the notion of place appears to result because it is not just a formal concept awaiting precise definition, but also a naïve and variable expression of [otherwise complex] geographical experience' (1986: 4). As was argued in Chapter 3, this definitional imprecision is productive to the extent that it is anti-reductive, resisting attempts to achieve succinct linguistic understanding, or to arrive at some kind of 'essence' of place. Rather, the pervasiveness of place and its plurality of forms means that 'it allows no hold', as Blanchot (1987: 14) might say, but its ubiquity and diffuseness also makes place most important as it informs and shapes lived existence. This position is one that will be both qualified and strengthened by the discussion to follow below.

However, the remaining three of the four issues listed above, present greater challenges for thinking about place in productive ways.

The first of these concerns the persistent conception of place as local and bounded. Some sense of boundary remains important in understandings of place. As Jeff Malpas (1999: 33) puts it, 'Fundamental to the idea of place would seem to be the idea of an open and yet bounded realm within which the things of the world can appear and within which events can "take place"'.

The difficulty that is posed by the idea of boundary in relation to place is that the emphasis in the above passage on ideas of 'openness' and 'event' are all too often neglected (or repressed). Instead, the idea of 'boundary' is taken in isolation and extrapolated to suggest that 'place' is (or has the potential to be) exclusionary. For instance, we see that the idea of place as 'closed' and therefore exclusive is strongly evident in the poststructuralist disavowal of 'community'. As was discussed above, the notion of community is resisted on the grounds that it inevitably leads to communion and fusion into oneness, thus becoming 'terroristic'. For this to work there is a requirement that, in part at least, 'both personal identity and a "place called home" have had to be conceptualised in a particular way—as singular and bounded' (Massey, 1992a: 15).

Of course, such a conception of place is not the sole preserve of poststructuralism. Rather, it is an association with a long history, particularly within the social sciences. Moreover, this association of place with the local and the exclusive lies at the heart of opposition from within certain strands of cybercultural criticism to the use of place and community metaphors in relation to CMC.

This is despite the fact that to conceive of place solely in this way is contestable. As Massey points out, it is a dubious formulation for two reasons. On the one hand, '"place and "community" have only rarely been coterminous' (Massey, 1991: 24). On the other hand:

Of course places can be home, but they do not have to be thought in that way, nor do they have to be places of nostalgia. [. . .] And what is more, each of these home-places is itself an equally complex product of the ever-shifting geography of social relations present and past. (Massey, 1992a: 15)

That is to say, place is never fixed, nor is it internally homogenous. Rather, place can best be understood as a bounded but 'open' and contested site, a 'complex product' of competing discourses, 'ever-shifting social relations', and internal (as well as external) 'events'. As Malpas puts it:

A characteristic feature of any 'place' [. . .] is that the elements within it are both evident only within the structure of the place, while that place is itself dependent upon the interconnectedness of the elements within it—as it is also dependent on its interconnection with other places—and, consequently, the idea of place is itself the idea of a structure that must resist any analysis that reduces it to a set of autonomous components. (Malpas, 1999: 39)

As for the counterposition of place and space, Massey argues that while many written accounts of place can in fact be quite varied in context and argument, there are nevertheless 'often shared undergirding assumptions', especially 'of place as closed, coherent, integrated as authentic, as "home", a secure retreat; of space as somehow originarily regionalised, as always-already divided up' (Massey, 2005: 6). 'More than that again', Massey contends:

They institute, implicitly but held within the very discourses that they mobilise, a counterposition, sometimes even a hostility, certainly an implicit imagination of different theoretical 'levels' (of the abstract versus the everyday, and so forth), between space on the one hand and place on the other. (2005: 6)

This understanding is also evident at a number of points in this book. It emerges, for instance, in the way that 'virtual space' is framed in opposition to 'real' space in some early accounts of the Internet and CMC, as well as in the belief in geographical 'liberation' and celebration of abstraction in the literature on 'cyberspace' (or 'neuromantic') architecture.

Massey resists this counterposition of place and space on the basis that it 'rests upon a problematical geographical imagination':

The couplets local/global and place/space do not map on to that of concrete/abstract. The global is just as concrete as is the local place. If space is really to be thought relationally then it is no more than the sum of our relations and interconnections, and the lack of them; it too is utterly 'concrete'. (Massey, 2005: 184)

Having provided this background to what Doreen Massey is working *against* in her critique of place and space, it is necessary now to shift to outlining what, in response to this, she is working *towards*.

In brief, Massey's project is to move beyond these restrictive formulations and develop what she terms 'an alternative approach to space' (2005: 9), 'an alternative positive understanding' (140) of space and place. This is based on three propositions regarding *space*, which are as follows:

1. 'That we recognise space as the product of interrelations; as constituted through interactions, from the immensity of the global to the intimately tiny.'
2. 'That we understand space as the sphere of the possibility of the existence of multiplicity in the sense of contemporaneous plurality; as the sphere in which distinct trajectories coexist; as the sphere therefore of coexisting heterogeneity. Without space, no multiplicity; without multiplicity, no space. If space is indeed the product of interrelations, then it must be predicated upon the existence of plurality. Multiplicity and space as co-constitutive.'
3. 'That we recognise space as always under construction. Precisely because space on this reading is a product of relations-between, relations which are necessarily embedded material practices which have to be carried out, it is always in the process of being made.' (Massey, 2005: 9)

As these propositions refer specifically to rethinking *space*, what, then, are their ramifications for how we might approach and understand *place*? The answer is that, for Massey, place is implicated fully in the same processes. She writes:

> What is special about place is not some romance of a pre-given collective identity or of the eternity of the hills. Rather, what is special about place is precisely that throwntogetherness [that is the result of contemporaneous plurality and coexisting heterogeneity], the unavoidable challenge of negotiating a here-and-now (itself drawing on a history and a geography of thens and theres); and a negotiation which must take place within and between both human and nonhuman. (2005: 140)

As she goes on to explain: 'This is the event of place in part in the simple sense of the coming together of the previously unrelated, a constellation of processes rather than a thing. This is place as open and as internally multiple' (140–141).

What is constituted here is quite a different conception of place and space from how these terms have sometimes been traditionally understood. And the implications of this alternative positive approach to place and space are specifically political. 'Both the romance of bounded place and the romance

of free flow', Massey writes, 'hinder serious address to the necessary negotiations of real politics' (175, see also, 183–184).

Herein rests the real force and significance of Massey's thinking on place
for this book. What a politics of place and the event of place demand, she
writes, is 'the ethics and the responsibility of facing up to the event; where
the situation is unprecedented and the future is open' (141). To conceive of
place in this way is to put 'on the agenda a different set of political questions' (141). In a crucial passage—one which reads as if aimed directly at
the poststructuralist critique of community and place—Massey states:

> There can be no assumption of pre-given coherence, or of community
> or collective identity. Rather the throwntogetherness of place demands
> negotiation. In sharp contrast to the view of place as settled and pre-
> given, with a coherence only to be disturbed by 'external' forces, places
> as presented here in a sense necessitate invention; they pose a challenge.
> They implicate us, perforce, in the lives of human others, and in our
> relations with nonhumans they ask how we shall respond to our tem-
> porary meeting-up with these particular rocks and stones and trees.
> They require that, in one way or another, we confront the challenge
> of the negotiation of multiplicity. The sheer fact of having to get on
> together; the fact that you cannot (even should you want to, and this
> itself should in no way be presumed) 'purify' spaces/places. (Massey,
> 2005: 141–142)

This passage is significant for several reasons. It acknowledges the
importance of difference and 'the negotiation of multiplicity'. But at the
same time, it does not position place as marginal or antithetical to difference, on the contrary, in fact. For Massey, place and space are central to
the social and central to the negotiation of difference. This is something
she has stressed on a number of different occasions, emphasising that 'the
very possibility of any serious recognition of multiplicity and difference
depends on a recognition of spatiality' (quoted in Rodgers, 2004: 283; see
also, Massey, 1992b: 79–84).

At this point, it is perhaps worth noting that Massey's conception of
place and space is not without criticism and the potential for misinterpre-
tation. For example, in her resistance to the place/space counterposition,
it can be unclear at times what precisely differentiates the two concepts,
or how each is defined. Going even further than this, Michael Hardt and
Antonio Negri (2000: 426, note 2) contend that, 'A notion of place that has
no boundaries empties the concept completely of its content'.

Both interpretations, Massey insists, misrepresent her work. For instance,
she argues that to refuse or challenge the counterposition of place and space
is not the same thing as collapsing one into the other, or inverting their
meanings, or denying any difference between them. On this point, and in
response to other criticisms and confusions, Massey is quite clear:

I must insist here, quite passionately, on one thing. This is not, as it is on occasions understood to be, a position which is hostile to place or working only for its dissolution into a wider space. Nor is it a deconstructive move, merely exposing an incoherence within an imagined essence (nor indeed is it proposing that what is at issue is purely within the discursive). It is an alternative positive understanding. This is certainly not to argue against 'the distinctiveness of the place-based' nor—and most particularly—is it to declare 'that there is nothing special about place after all'. Quite to the contrary. (Massey, 2005: 140)

Massey's 'alternative positive understanding' is to emphasise the 'event of place' (2005: 140). Giving due emphasis to the 'event of place' forms a key aspect of this, the third part of the three-part proposal outlined in this chapter. The argument here is that Massey's relational understanding of the event or politics of place has manifold implications and value for thinking about teletechnologies, community and place. However, prior to detailing the fuller implications and value of this understanding of the event of place for the material examined in this book, it is necessary to turn briefly to the last of the issues noted at the beginning of this section. This is the idea that place is often regarded as extrinsic or subsidiary to human social experience.

A crucial insight to be drawn from Massey's 'alternative positive understanding of place' is her general belief in the ongoing importance of place and space for understanding, negotiating, and constructing who we are in social terms. As she puts it, 'There is no getting away from the fact that the social is inexorably also spatial' (1992b: 80).

Jeff Malpas goes much further than this in his engagement with the idea of place. He argues that 'it is within the structure of place that the very possibility of the social arises' (1999: 36).[4] Even more fundamentally still, Malpas proposes that in fact, 'the structure of subjectivity is given in and through the structure of place' (36). As he puts it at an earlier point in the same text, 'The very possibility of the appearance of things—of objects, of self, and of others—is possible only within the all-embracing compass of place' (15).

What is significant about the above arguments is that to propose that place is instrinsic to human subjectivity and human experience (as opposed to supplementary or extrinsic to these things) is, in turn, to effectively argue, as Massey does, that it is 'open to politics' and open to difference (Massey, 1992b: 82). For example, Malpas's argument that the structure of subjectivity is given in and through the structure of place is important in light of Honi Fern Haber's and Todd May's critique of poststructuralism detailed above. Central to this critique is their argument for the recuperation and ongoing importance of the notion of the subject, which they assert is crucial to the viability and sustenance of any effective 'politics of difference'.

Meanwhile, for Massey, the argument that place is open to politics comes from her belief 'that the identity of any place, including that place called home, is in one sense for ever open to contestation'; the identities of places 'are inevitably unfixed' (1992a: 13). This is to say, among other things, that places, like actualities (indeed, *places as actualities*), are open to change and open to the future.

This position shares with Derrida an explicit concern for ethico-political responsibility, particularly with respect to how places are made and the various 'power-geometries' that are at play (see Massey, 2005: 141, 154, 180).[5]

It also highlights the importance of thinking place through, and in relation to, what she terms a 'politics of connectivity'. Massey writes:

> A relational politics of place, then, involves both the inevitable negotiations presented by throwntogetherness and a politics of the terms of openness and closure. But a global sense of places evokes another geography of politics too: their construction. It raises the question of a politics of connectivity. (Massey, 2005: 181)

While Massey is referring to connectivity in a number of different senses here (181ff), one of these senses does include networked information and communications technologies and the role they play in the construction of a relational politics of place.

Some of the implications this understanding has for communications have been teased out in an essay on the work of Massey by media critic Jayne Rodgers. Rodgers (2004) argues that Massey seeks 'greater recognition of the intersecting, dissecting, crosscutting relations that are constitutive of space' (288) and that her emphasis on the 'genuine and potential multiplicities of the spatial' (287) is significant for thinking about teletechnologies, especially networked computing. These multiplicities represent many different ways of conceiving, experiencing, and constructing space (288). To cite the example of cyberspace, Rodgers suggests that there are 'millions of on-line and offline spaces, frequently intersecting and each having an impact on both the user and non-user in how space is constructed and how it evolves (288).

This understanding is implicit in Lance Strate's approach to cyberspace as inherently plural, and is why Strate's account of the multilayered structure of cyberspace holds merit (see Chapter 3). It is also implicit in Cedric Price and Joan Littlewood's Fun Palace project (see Chapter 4). In overall terms, the above passage also captures what it is that I am arguing in this book is crucial if we are to begin to more fully understand the complex interactions and interconnections of teletechnologies, community and place.

When place is understood as multiple and open to ongoing contestation, then any residual maintenance of (and belief in) a virtual/real dichotomy collapses completely. So, too, does any conception of cyberspace as a singular rather than plural phenomenon. Rather, what emerges instead, as Rodgers (2004) and others (Ek, 2006) make clear, is a far more complicated picture of technosocial and techno-*spatial* (and -*'placial'*) interaction and engagement.

To reconceive of place in the way that has been proposed above—as open, relational, and as intrinsic to who we are as human subjects—has manifold implications for the material examined in this book.

For example, the poststructuralist critique of community proceeds in part from the assumption that community and place have historically been bound to one another and that this bond is problematic. The alternative model of community developed in this chapter, and the conception of an open, relational understanding of place that is developed by Massey (and also to an extent by Malpas), combine to complicate this assumed bond. Put another way, there will always be points of intersection between ideas and instances of place and community, but these points of intersection should not be assumed to be inseparable or restrictive.

An understanding of place as open and relational also holds implications for metaphor use. For example, Malpas argues that the complexity of place as a concept complicates many of the prevailing assumptions that underpin the usage of place-based metaphors. He writes, 'The claim that "place" can remain only a metaphor in this context is simply a reassertion of a particular and fairly narrow view of the nature of place—a reassertion that seems to ill-accord with the complex character of the concept' (Malpas, 1999: 38). In writing on CMC, this 'narrow view' is an often-assumed understanding of place as local and bounded. It is also on the basis of this understanding that these metaphors are resisted in the critical literature on CMC and the Internet. Yet, the understanding of place advocated in this chapter provides a very different basis for possible metaphor creation and tantalising possibilities. While it remains in the realm of speculation what the broader impact of such metaphors might be for cybercultural research, two preliminary outcomes are likely. First, to conceive of place as open and relational is likely to bring new vitality to place-based metaphors. Secondly, metaphors developed from this conception of place would be even more pertinent as a critical tool for making sense of teletechnologies, given that these technologies (as the prefix 'tele' suggests) are fundamentally technologies of distance, relationality, and connectivity.

In relation to the material of Chapters 4 and 5, an understanding of place as open, relational, and interconnected as advocated here is at significant odds with how place is framed and understood in the writing of architectural computing, where place is either a peripheral concern, or it is dismissed altogether. For example, in early writing on architectural computing, place is of reduced importance relative to other considerations. Whereas in later writing on architectural computing and cyberspace, place is seen as antithetical to vision of a dematerialised, transmissible architecture. This particular vision of architectural and bodily transcendence is a fantasy constructed around a whole series of assumptions, not the least of these being a very narrow conception of place as local, bound and untouched by teletechnological and other forms of connectivity. A key exception to the above is the work of Cedric Price. All of Price's architectural explorations discussed in this book display an implicit awareness of the sort of 'politics of connectivity' of which Massey writes, as well as an awareness of place as open and relational.

Lastly, in relation to the material of the previous chapter, it would seem that research concerned with the domestication of technology and networked mobility is far more acutely attuned to an understanding of place as open and relational. In fact, one of the key arguments of Chapter 6 is that research into mobile technologies suggests that the social uses of these technologies both reinforces and intensifies an understanding of place as open and relational and complex in how it is experienced (see also, Wilken, 2008; Richardson and Wilken, 2009). Moreover, of all the material examined in this book, it is this research into mobile telephony that comes closest to (without perhaps fully realising) Malpas's (1999: 36) argument that the very possibility of the social and the structure of subjectivity are given in and through the structure of place.

LINES OF FLIGHT

> 'To make an end is to make a beginning. The end is where we start from.'
>
> —T. S. Eliot (1943: 14)

In developing the above three-part proposal, the task has been to present productive theoretical possibilities for making sense of the complex interactions and interconnections between teletechnologies, place, and community. In combination, these arguments concerning actuality and teletechnology, community, and place, are significant in that they bring new inflections and angles of approach to these crucial issues and ideas.

From an interdisciplinary perspective, the framework of this proposal provides a productive initial step towards the formulation of understandings of and approaches to community and place within a 'shared language' that is nevertheless attentive to and respectful of differing angles of disciplinary approach to these concepts.

In overall terms, the proposal highlights the ongoing importance of ideas of place and community for understanding and engaging with teletechnologies. It is significant in that it enriches conceptual understandings of place and community as ideas, bringing new inflections and angles of critical approach to these ideas and the issues that flow from them. It also highlights the need for continuing, careful, critical engagement with these concepts that is ethically engaged and anti-reductionist. Furthermore, its aim is that these alternative angles of approach and understanding will shift the terrain of the debate about teletechnologies, place, and community, by providing a systematic and productive point of departure for engaging with and making sense of the transformations—both social and spatial—that are being wrought by human-technology interaction, technological infrastructure, innovation, and design.

However, it is by no means suggested that this proposal—which combines Derrida's understanding of 'artifactuality' and 'actuvirtuality', with reworked understandings of community and place—resolves or fully responds to *all* issues pertaining to the intersections and interactions of teletechnologies, place, and community. In other words, it does not represent nor aim to provide an overarching or grand ('totalising') theory of the techno-socio-spatial.

Rather, it should be emphasised that this three-part proposition for a theoretical shift is by no means complete or fully resolved. On the contrary, significant further work is required in examining and establishing just how the three elements or facets of this framework hold together and interact, as well as in identifying at what points and in what circumstances the three elements come into conflict or friction with one another. Reflecting on some of these issues below serves as a valuable counterpoint to the emphasis given elsewhere in this book on media historiography.

The three-part proposal developed here would benefit, for instance, from further development and testing through application to, and study of, concrete examples: specific virtual communities, actual urban architectural projects, planning developments, 'community-building' projects (especially in the fields of urban and community informatics), and so on. Explorations of this sort would prove productive in testing the usefulness and limits of the three-part proposal or framework; it is also likely to yield further insights into the intersections and interconnections of teletechnology use, and ideas of place and community that would bring further refinement to present understanding of these facets and their interrelationships.

In addition, the proposal could well contribute to larger debates concerned with the ways that teletechnologies, place, and community function as facilitating and enabling concepts, and, on the other hand, as constraining and repressive concepts.

In terms of what this trio of concepts might enable, there is mounting evidence of their potential for mass organisation and collective action (Rheingold, 2008, 2002; Castells et al., 2007: 185–213; Warf and Grimes, 1997). For example, the early use of the Internet by the Zapatista movement in south-central Mexico provides some support for the view that digital practices—especially when mobilised in conjunction with ideas of place and community—can serve as a liberating force (Froehling, 1997; Chabrán and Salinas, 2004: 322). There have also been a number of examples of the use of teletechnologies (such as the Internet and mobile teletechnologies) to intervene in the political process and mobilise people in various forms of public protest and collective action. Rheingold (2008) furnishes a litany of examples, which range from the by now well-known EDSA-2 'people power' demonstrations in the Philippines and various grassroots protests in China, to election monitoring in Africa and non-violent anti-government rallies in Spain. A particularly striking example is that from the former Federal Republic of Yugoslavia, where the student-coordinated group Otpor combined clever non-violent political protest with teletechnological savvy to orchestrate the ousting of the country's leader, Slobodan Milosevic (*Bringing Down a Dictator*, 2001), tactics which have since been replicated elsewhere. Supplementing the above examples are less overtly politically motivated forms of mass organisation and collective action, such as the once popular playfully mischievous phenomenon of 'flashmobbing' (Marchbank, 2004; Nicholson, 2005). This can take a range of forms, but generally involves temporary gatherings in public places to perform coordinated yet fleeting acts of absurdity which are organised by Internet or mobile phone—such as, to cite one example, meeting at a particular time and place to point yellow rubber-glove-clad hands in the sky, and then quickly dispersing (Marchbank, 2004).

These are just a few instances where the combination of teletechnologies and ideas of place and community are seen as enabling mechanisms.

Of course, these same tools and concepts can also be combined and utilised to promote social exclusion and fuel ethnic and religious divisions. A vivid Australian example of this is the alleged use of SMS to coordinate 'race riots' on Sydney's beaches in December 2005 (Goggin, 2006). This particular example highlights the ambivalences which attend the social uses of teletechnologies, where these uses are far from straightforward or uncomplicated. Rather, and as noted above, they emerge from and within complex broader processes of social, political, and technological change.

In terms of what else these three concepts repress and constrain, the list is potentially a long one. For example, the issue of difference and otherness has been touched on in Chapter 2 in relation to the notion of community, and elsewhere in this book (see also, Gandhi, 2006), as has the potentially constraining function of metaphor for understanding teletechnologies. For example, in Chapter 3 it was noted that many critics, especially within the area of media law, see place-based metaphors as potentially constraining because they can lead to the imposition of private property conceptions onto teletechnologies such as the Internet which have the potential to influence Internet governance. It was also argued in that chapter that place-based metaphors can be limiting because place is often conceived in a particular way (as local and bounded), and this understanding leads to the rejection of place as a constraint to be overcome and therefore to a denial of place. These particular considerations of facilitation and constraint also dovetail with broader spatio-cultural and political questions of 'access' (such as long-standing 'digital divide' arguments), as well as with issues concerning what David Sibley collectively calls 'geographies of exclusion' (Sibley, 1995; Marx, 1997). Any consideration of what this trio of concepts (but particularly teletechnologies) might facilitate and enable, thus must also recognise the ambivalences of these concepts, and must be situated within a wider framework of socio-political, socio-spatial, and technological change.

In a related vein, there is also the question of why this particular trio of concepts—teletechnology, place, community—bobs up and down in time and across cultures, with each (or all) of them coming into focus at certain times and passing out of focus at others. For instance, why is it that the notion of 'community' and especially 'virtual community' received such heavy exposure during the early-to-mid-1990s, especially in the United States, in the literature on computer-mediated communication, only to be displaced or supplanted by other concerns, debates, and technological developments in the early-to-mid-2000s? A part explanation for this waxing and waning process, especially in relation to technologies, can be found in Andrew Calcutt's observation that:

> current trends in society are usually described as if they are effects which have been caused by the new technology. [. . . However,] technology can only be developed and applied in accordance with the social context from which it is derived. Technology which does not correspond to the mood of the times is likely to be discarded. (Calcutt, 1999: ix)

Calcutt's point can be further illustrated via the example of architecture's interest in cyberspace. As discussed in Chapter 5, the idea of cyberspace held special poignancy within architectural discourse in the early-to-mid-1990s. Yet, as one critic notes, 'today, hardly anyone seems content with that notion' (McCullough, 2004: ix). There are two possible reasons for this shift: what McCullough describes as 'a turn from the fast and far-reaching to the close and slow', and an architectural 'paradigm shift from building virtual worlds toward embedding information technology into the ambient social complexities of the physical world' (ix). The first of these reasons is, in effect, an acknowledgement that 'place matters even within the space of flows' (quoted in Sinclair, 2003: 217), as do the temporal and other rhythms that are unique to each place (Lefebvre, 2004). The second recognises ongoing shifts in the focus of teletechnology innovation, from stationary computing (the desktop PC), to ubiquitous, pervasive, and wearable computing, and an overall emphasis on transportability and increased miniaturisation and the spatio-cultural transformations wrought by these developments (Galloway, 2004; Galloway and Ward, 2006; Andrejevic, 2005).

Accounting fully for these and other shifts also requires engagement with an array of broader issues and developments that straddle political economy, socio-cultural and socio-political change, scientific discovery and innovation, production and consumption patterns and processes, and so forth. Here the above three-part proposal can provide a constructive point of entry into, and potential for further refinement of, larger debates about networked society and globalisation. This is because the trio of concepts that form the focus of this book are all 'constitutive elements of broader processes of globalization, modernization, industrialization, economic restructuring and cultural change' (Graham, 2004: 18). The intention of the earlier tripartite proposal is to provide a series of initial and productive steps towards the development of a fuller, more unified, and theoretically coherent response to the issues raised throughout this book. Even so, it should be noted that this proposal works at a particular scale; a different scale with a different set of priorities and theoretical inflections and focus might require a different set of considerations and propositions. For example, such as reframing this examination of teletechnologies, place, and community to focus on how they intersect with broader considerations of globalisation, as well as more specific considerations within these relating to diasporic communities, larger patterns of transnational migration, and global mobility. Each of these will be touched on below.

In terms of the first of these, there is scope for thinking through the implications of the earlier proposal in relation to how social, cultural, artistic, scientific, economic, ethnic, and other diasporas are beginning to reconstitute and reconfigure, and redefine or reassert, place, community, and identity. To use the example of ethnic diasporas, this work is particularly important insofar as place—and especially the idea of 'a place called home', as Doreen Massey (1992a) puts it—is a complex and contested notion that is in constant negotiation for many diasporic communities. For example, in her study of Vietnamese women migrants to Australia, Nathalie Nguyen observes that for these women, as for many other migrants and displaced persons, their country of origin 'is no longer "home"', yet, at the same time, their country of destination is equally 'not fully "home"' (Nguyen, 2005: 148). As was explored in the previous chapter, as it is in the work of other critics (for example, Morley, 2000), these tensions are also closely tied to questions of teletechnology use.

At the same time as there is this somewhat 'liminal' experience of place, media flows are becoming increasingly complex and are not only breaking down traditional centre-periphery distinctions (such as West-East, first world-third world, and city-country), but are also 'beginning to define new kinds of world region' (Sinclair and Cunningham, 2000: 2). These include 'geolinguistic regions', which are described as 'regions across which linguistic and cultural similarities are at least as important as geographical proximity has been in forming world regions in the past' (2 ; see also Wilken and Sinclair, 2011). Moreover, in addition to complex patterns of media consumption (Karim, 2003), it is argued that 'the media space of a diaspora' tends to be of a geolinguistic regional kind, 'to the extent that it is spread throughout several of the national markets which have been the territorial unit for international media distribution in the past' (Sinclair and Cunningham, 2000: 2; also, Gillespie, 2000). Suffice it to say, these developments and transformations pose important and complex questions for understanding the intersections and interconnections between teletechnology use, place, and community (see Poster, 1998).

The earlier proposal also feeds into, and in turn would benefit from, consideration of how the same trio of concepts might manifest themselves in different socio-cultural, political, and national contexts. As mentioned in the Introduction, the focus of this book has largely been on European and Anglo-American engagements with teletechnologies and ideas of place and technologies. What remains to be examined is how these concepts circulate in the literature of other cultures (Goggin and McLelland, 2009b), and how they manifest themselves and are utilised in 'non-Western' or 'developing' countries. How, for example, are the interactions between teletechnologies, place, and community manifest in diverse socio-cultural and political contexts, such as North Korea, or China, or Cuba, or sub-Saharan Africa? How do these differ from England, Europe, North America, and Australia? Moreover, how are these interactions manifest in the equally complex socio-political contexts of Mexican-American *barrios*, for example, or Brazilian *favelas*? While some significant work has already been done in this area (Appadurai, 1990; Castells, 1996; Davis, 2006), filling in the detail of this broader, global picture remains a long-term undertaking and ongoing task.

In addition, the increasing global ubiquity of mobile telephony (Goggin, 2011; Castells, et al., 2007) and networked telecommunications also needs to be accounted for in these considerations of teletechnologies, place, and community. For instance, since 2001 there has been a reported significant levelling off in the number of Americans who use the Internet (Rainie and Bell, 2004), while global mobile use continues to grow. What is remarkable about this technological shift is that, unlike past Internet access, mobile use cuts across class, racial, and socio-economic boundaries. This is partly because mobile phones are now 'well within reach of all but the poorest' (Philipson, 2006). Moreover, while pure growth in new connections does not necessarily equate with equal global distribution and access, it is significant that a quarter of the growth in new mobile connections is coming from two countries, China and India, while 'most of the rest is coming from other developing countries, with Pakistan, Bangladesh, Brazil, Indonesia and Nigeria all in the top 10' (Philipson, 2006). What this shows (yet again) is that teletechnologies—especially mobile teletechnologies—cannot be ignored, and that detailed and sustained critical engagement with these technologies is crucial, especially given the fact that mobile technologies (as discussed in the previous chapter) impact on and shape understandings and experiences of place and community in significant ways.

Finally, and to build from this situation, there is the need to also think through and account for complex intersections between mobile technology use and broader geopolitical questions of place, community, and especially mobility and migration. David Morley advises that any analysis of technology and questions of place 'must be sensitive both to [what Foucault terms] the "grand strategies of geopolitics" and the "little tactics of the habitat"', where the 'interlinked processes of globalisation and domestication [. . .] bring together micro and macro issues' (2003: 437).

This brings me back, in conclusion, to Derrida's account (discussed briefly in Chapter 6) of the return toward home, the 'at-home', in both its benign domestic sense and more troubling nationalistic sense. It is the former sense which would seem to drive present interest in and uptake of mobile telephony technologies. As Morley puts it, these technologies are often understood 'as "imperfect instruments, by which people try [. . .] to maintain some sense of security and location" amidst a culture of flow and deterritorialization' (2003: 453). The persistence of place in the face of CMC and networked mobility 'seems to suggest a continuing desire to reterritorialize the uncertainty of location inherent in online worlds' and networked telecommunications (440). But it is worth remembering that this is not unconnected from the somewhat darker nationalistic desire for the 'at-home', which, as Derrida explains, is motived (among other reasons) by the perceived threat that is posed by the 'mobility' of the immigrant 'Other'. Future research in this area would do well to remember this, by remaining sensitive to both the micro-scale of (largely localised) experiences of networked mobility (Urry, 2002), and the macro-scale of global geopolitical transformations (Urry, 2000a, 2000b), the micro-politics of mobile, teletechnologically equipped bodies in transit through place/s, and the macro-scale geopolitics of voluntary and forced migration and displacement.

To close, this book has examined just some of the relationships between teletechnologies, place, and community. It has done this by exploring how notions of place and community circulate in the literature on teletechnologies and place-making. Above all else, its central themes have been that, while teletechnologies cannot be ignored, neither can place and community. These last two are persistent concepts that continue to inform and shape, and be shaped by, our engagement with teletechnologies. In combination with teletechnologies, ideas of place and community continue to be vital in helping us think about and experience who we are, where we are, and the ways that we interact and relate with one another.

Notes

NOTES TO THE INTRODUCTION

1. For example, as Nick Couldry and Anna McCarthy write, 'one could almost call media and space the *obverse* of each other, necessarily connected but, as Foucault says, "irreducible to one another"' (2004: 1).
2. Thus, 'interdisciplinarity' is here taken to include the transdisciplinary. That is to say, as it is used here, the term is taken to incorporate that which is between, among [*inter-*] disciplines, as well as that which is across, beyond [*trans-*] disciplines.

NOTES TO CHAPTER 1

1. For example, David Whittle (1997: 6) remains relatively true to Gibson's conception, but extends his definition beyond 'the conceptual world of networked interactions' to include 'the state of mind shared by people communicating [via these networked interactions]'. John Parry Barlow once famously quipped that cyberspace is 'that space you are when you are talking on the telephone' (quoted in Whittle, 1997: 6). While for Howard Rheingold (1994: 5), to cite one of a number of definitions he offers, cyberspace 'is the name some people use for the conceptual space where words, human relationships, data, wealth, and power are manifested by people using CMC [computer-mediated communications] technology'. Lastly, Lance Strate (1999) conceives of cyberspace as a multivalent term which carries at least three levels of meaning.
2. To emphasise this understanding of cyberspace as an alphabetic space, Tofts contracts the word cyberspace to become 'cspace' (pronounced 'space'). The idea is that the subtleties of this contraction can only be grasped through reading, not speech. As he explains elsewhere (Tofts, 2004a), this contraction 'refers to the modification of consciousness through the interiorization of the phonetic alphabet, to a mediated apprehension of the world through the abstract interface of acoustic and visual space'.
3. For concerted attempts to correct aspects of this historical amnesia, see Milne (2010), Gitelman (2008), Chun and Keenan (2006), Gitelman and Pingree (2003), Tofts, Jonson and Cavallaro (2002), Sconce (2000), Bolter and Grusin (1999), Tofts and McKeich (1998), and Marvin (1988).
4. Indeed, as management scholars have noted, 'of the many ideas that have entered the business world by way of the Internet, few have proved more potent than "online community"' (Williams and Cothrel, 2000: 81).

5. In particular, see Dyson (1997: esp. 31–53), Bollier (1995), and OECD (1998). In all these sources the community metaphor is unproblematically extended to CMC. Indeed, the comparison is taken to be almost self-evident.

6. This formulation of 'imagined' community owes a considerable, albeit problematic, debt to the work of Benedict Anderson (1983). For a sharp critique of the appropriation and transfer of Anderson's notion of imagined community to descriptions of virtual community, see Brabazon (2001).

7. For a discussion of what has been described as 'the struggle for possession of the community metaphor', see Watson (1997: 120ff). According to some critics, even the first term in the metaphor—the word 'virtual'—is an extraneous inclusion. For example, it is with a degree of astonishment that the Norwegian critic Espen Aarseth observes: 'even among social researchers who observe and experiment with the MUDs, such text-based social interactions are sometimes referred to as "virtual communities", as if real community cannot take place in digital, nonlocal communication but needs a physical, tangible space to exist properly' (1997: 146).

8. Feenberg and Bakardjieva (2004: 41) remark that, 'It is a good time to remind ourselves that online groups are indeed a qualitatively new medium. We are far from a final understanding of them. Rather, we are at the point where systematic research on them can begin to bear fruit. [. . .] Research can identify, describe and understand the specific forms of social life within computer-supported environments and the related benefits, drawbacks and consequences for participants, culture and society.'

9. For a summary of these debates, see Weimann (2000: 343–352), and for further discussion of the social theory implications of this, see Lyon (1997: 23–37).

10. As Philip Agre (1998: 70) notes, 'Everybody's daily life will include a whole ecology of media; some of these will be voluntarily chosen, and others will be inescapable parts of life in public spaces and the workplace'. For a similar, but somewhat more cumbersome, model to that proposed by Mitchell, see James and Carkreek (1997).

11. In a study of North American academics who were participants in two scholarly networks, Koku et al. (2001: 1752) discovered that, 'Frequent contact on the Internet is a complement for frequent face-to-face contact, not a substitute for it. The more scholarly relations network members have, the more frequently they communicate and the more media they use to communicate'. In a similar vein, Ling (2000: 61) argues that relationships on the Internet are 'slower to develop and necessarily migrate over to other forms of communication including face to face interaction'.

12. 'The etymology of the Greek *ecstasis*,' Neil Spiller explains, 'is concerned with the notion of disjuncture: to be beside yourself, out of your face or not full of your skin; that is, to occupy a non-body-centric space. This derangement of the body and "soul" is an escape from their usual earthly relationship. The dislocation of body and soul encourages a perceptually different order of experience, a readjustment of the normal mental and visual perceptual parallax.' (Spiller, 1998a: 61)

13. As Gye (2004: 33) observes, 'There is a persistent trope of disembodiment that frames many investigations into the impact of new media technologies on culture. [. . .] Despite the aching necks and backs and chronic RSI suffered by those who engage heavily with computing technologies, this trope persistently resurfaces'.

14. At an earlier point in the same text Spiller goes even further: 'It might now be possible for the self to discard the oldest prosthetic of all, the body. With the cyberspace prosthetic it may be possible to escape the "meat" for good, with no harm caused.' (Spiller, 1998a: 34)

15. For further, useful critical discussions of disembodiment, see in particular, Vasseleu (1997), Franck (1995, 1998), and Argyle and Shields (1996).
16. In her own work, Milne (2010) explores in detail the construction and persistence of presence, identity, and embodiment in email exchange.
17. Milne (2010, 2007) argues a similar point in relation to the construction of presence and identity in the Cybermind email list.
18. This is supported elsewhere. For example, in her study of the Cybercity virtual community, Carter (2005) finds that 'human relationships in cyberspace are formed and maintained in similar ways to those in wider society [and] rather than being exotic and removed from real life, they are actually being assimilated into everyday life' (148 and 163–165). Likewise, in a study of the LambdaMOO, Schiano and White (1998: 359) observed that 'major patterns of behavior in this text-based virtual world do not depart radically from those in "real life"'. Similar conclusions are drawn in Lori Kendall's (2002) study of masculinities and relationships online.

NOTES TO CHAPTER 2

1. For Bauman (2001: 144), this is because community 'remains stubbornly missing, eludes our grasp or keeps falling apart'.
2. This view owes a considerable debt to the work of German sociologist Ferdinand Tönnies, which will be discussed later in this chapter.
3. As Ben Highmore (2002: 106) puts it, 'the danger of Fascism could be seen precisely in the spectacular lure of the "mass" coming together in an erasure of difference'.
4. These concerns also coincide and intersect with feminist critiques of community, such as by Weiss (1995), Weiss and Friedman (1995), and Young (1995).
5. On the influences informing Tönnies's thinking, see Donini and Novack (1982), Sorokin (1963), and McKinney and Loomis (1963). On the influences informing Schmalenbach's thinking, see Lüschen and Stone (1977).
6. For a fuller discussion of Tönnies work, and how his ideas can be read against those of other key thinkers in the sociological tradition, especially Weber, see Loomis (1963) and Cahnman (1995).
7. As Hetherington (1994: 6) explains, there 'was also a critique of Spengler and more significantly an attempt to apply a phenomenological approach, influenced by Husserl's practice of categorical intuition, to the Simmelian inspired study of social forms'. As Hetherington further notes, 'The essay also used the concept in a critique of Weber's theory of the routinization of charisma and of his fourfold typology of social action' (6). Hetherington's essay also offers an account of the reception by English-language social science scholars of Schmalenbach's concept (6).
8. Although it does receive some support from David Bell's (2001: 106) tale of students' first-hand and ambivalent accounts of *Gemeinschaft*-like communities.
9. Something of the complexity of (and problems associated with) the term within German—and wider European—cultural and political history is discernible in Günter Grass's (1979) proto-surrealist novel, *The Flounder*. This complex history reveals the difficulties associated with disentangling the concept of the Bund for extrapolation to other contexts, such as virtual community formation.
10. Hetherington suggests this at least partially explains the poor reception of Schmalenbach's concept in the succeeding years, but adds that 'it was the

Bund as an idealized form of sociation found in the youth movement that provided one major source of inspiration for Schmalenbach's sociological conceptual formulation and not this later manipulation of the idea of the Bund' (Hetherington, 1998: 87).

11. This understanding of the term 'community' as 'local community' is quite common in the architectural and urban planning literature. A particularly clear example can be found in the work of Serge Chermayeff and his various co-authors. In this work, the word community tends to refer to 'local community', and even that meaning is not spelled out, but rather it is assumed (see, Chermayeff and Alexander, 1963; Chermayeff and Tzonis, 1971).

12. One plausible reason that the second approach became more common than the first, particularly around the time Hillery was writing, has to do with the increasing methodological difficulties associated with applying a criterion such as 'area', which appears well matched to a rural setting, to the study of large urban and even suburban communities (see Hillery, 1955: 119).

13. Even in everyday local communities, the interactions between community and place are far from clear cut and are often complex, unstable, and in negotiation. For example, one ethnographic fieldwork study in a UK suburb discovered, 'via the incident of the search for a lost cat, how everyday talk formulates places and is formulated by its location in the ongoing occasioned activities of neighbours' (Laurier, Whyte and Buckner, 2002: 346). Keleman and Smith (2001) make a related point, when they remark that 'the construction of any form of human association is thoroughly sociological, in that it is dependent on mundane experiences which shape both the emotions and thought of human beings' (378).

14. Of course, Nancy and Derrida were not the first to identify the problems associated with such an understanding of community. As Delanty notes, as early as 1922 these were also noted by the German philosopher Helmut Plessner in his 1924 book, not translated into English until 1999, *The Limits of Community*. Here, Delanty explains, Plessner 'presented a major critique of the idea of community' in which he argued that 'community was an overvalued ideal and contained a latent authoritarianism' (Delanty, 2003: 22).

15. Finitude, as Nancy develops it in relation to the idea of a 'workless' community, is a complex notion which holds a central place in *The Inoperative Community* and in Maurice Blanchot's (1988) response to Nancy's text, *The Unavowable Community*. This complexity precludes a full explication here. Suffice it to note, as Pierre Joris explains in his translator's preface to Blanchot's text, that the notion of finitude is most strong in Bataille's association with the mysterious 'Acéphale' group, a social organisation with both a public presence and private aspect. The public presence was an eponymous magazine published four times between 1936 and 1939. The private aspect of the group involved 'meeting around trees struck by lightning' and talk of human sacrifice. The latter activity was never carried out, apparently due to Bataille's disputation on the basis of a logic which claimed that both consenting victim and executioner must perish simultaneously in order to achieve the desired goal of a 'headless' community (which is what 'acéphale' translates as) (see Joris, 1988: 50, note 2).

16. Although, as Schmalenbach writes, 'Even consanguinity does not generate social relations unless a commonality is *recognized* "by the persons concerned"' (1961: 331—emphasis added). In respect to this passage, Rick Parrish (2002: 263) argues that Schmalenbach's point is that, 'community actually precedes its recognition' and that recognition 'is the product of community, not its genesis'.

17. This 'disconcerting task' of 'enduring the affect that feels like a strange self' in order to 'expose oneself to others' arguably forms the closest point of convergence (if not contact) between Nancy's project of rethinking—or thinking at the limit of community—and Alphonso Lingis' (1994) meditation on community and otherness (see Wilken, 2010: 459–465).

18. The sequence is significant here. That a critique of community, and all the social and political meanings and implications that are carried by this term, is to be found in the so-called 'early' works of Derrida is important in that it counters a persistent and often strident criticism of Derrida's thinking in particular and poststructuralist deconstruction in general as 'apolitical'. It is also an issue that Derrida himself has responded to: 'Perhaps this brings us back to a more philosophical order of the response, [. . .] of the theme of difference, which has often been accused of privileging delay, neutralization, suspension and, consequently, of straying too far from the urgency of the present, particularly its ethical or political urgency. I have never understood there to be an opposition between urgency and difference. Dare I say, *on the contrary?*' (Derrida and Stiegler, 2002: 10; see also Lucy, 1995, 1997).

19. A somewhat similar argument, or so it would seem, is put forward by Zygmunt Bauman (quoted in Delanty, 2003: 119) who 'argues for a postmodern ethics based on individual autonomy and in which the exclusion of the other is not the price to be paid for the identity of the self'. This quote also gestures towards one problematic aspect of Corlett's reading of Derrida: the tendency to align him with postmodernism. The reasons this should be understood as an erroneous alignment are explained by Lucy (1997).

20. Derrida also notes in this passage the possibilities that this understanding of gift-giving and indeterminacy might hold for sexual identity (Derrida, 1987: 199).

21. Note that, here, the term 'singularity' is meant in a non-Nancean, pejorative sense, as a desire for national identity.

22. Van Den Abbeele argues that, in Nancy's case, he effectively disconnects 'the assumed immanence of communal identities to demarcated geographical spaces in the form of towns, lands or nations' (Van Den Abbeele, 1997: 15). 'In its most vulgar formation', Van Den Abbeele goes on to state, 'this relation appears of course as the nationalist ideology of blood and soil' (15). This separation should be understood as a desire to separate place from a *particular conception* of restrictive unity, not as a critique of place per se. Moreover, this separation can also be read as a perpetuation of a more general tendency, in the historical development of community thought, which was detailed earlier in this chapter, that sees area as one of the less significant features of conceptions of community.

23. As Hannah Arendt (1973) puts it, 'plurality is the condition of human action because we are all the same, that is, human, in such a way that nobody is ever the same as anyone else who ever lived, lives or will live' (8).

NOTES TO CHAPTER 3

1. And this is perhaps why Casey offers no concise definition of place. His suggestion seems to be that we reach an understanding of place only by taking a circuitous route: by studying 'the perplexing phenomenon of displacement, rampant throughout human history and especially evident at the present historical moment, only in relation to an abiding implacement' (1993: xiv).

2. Castells defines the 'space of places' as the 'the historically rooted spatial organization of our common experience' (1996: 378). By way of contrast,

he defines the 'space of flows' as a series of transformations where 'society is constructed around flows: flows of capital, flows of information, flows of technology, flows of organizational interaction, flows of images, sounds and symbols. Flows are not just one element of the social organization: they are the expression of processes *dominating* our economic, political, and symbolic life' (412—original emphasis). Or as Derrida puts it, 'the border is no longer the border, images are coming and going through customs, the link between the political and the local, the *topopolitical*, is as it were *dislocated*' (Derrid and Stiegler, 2002: 57—original emphasis).

3. The impact of mobiles on experiences and understandings of place will be explored in Chapter 6.

4. As one commentator explains, for Derrida '"metaphysics" does not simply equal "philosophy", if "philosophy" is understood to be simply what is taught as such institutionally' (Lucy, 1995: 66). Rather, 'Derrida's "metaphysics" approximates rather to something more like the space in which ideas (or concepts) are able to be taught' (66). More precisely, Spivak (1997: xxi) writes that, 'Derrida uses the word "metaphysics" very simply as shorthand for any science of presence'. In other words, the term 'metaphysics' is used by Derrida to designate the history of a way of thinking that 'requires us not to think of the "essence" of truth as a question that needs to be thought through, but rather as the fundamental ground or necessary origin of thought in general' (Lucy, 2004: 240). In key respects, this accords with Simon Blackburn's (1996: 240) rendering of metaphysics as that which 'tends to become concerned [. . .] with the presuppositions of scientific thought, or of thought in general'. In the present section, 'philosophy' and 'metaphysics' are used interchangeably, and both terms carry the above understandings.

5. Paul Ricoeur (1977) takes a very different view. Unlike Derrida, Ricoeur is not suspicious of metaphysics and instead maintains a belief in its ability to reveal truth. This faith is based on the human capacity to coin new metaphor. This is important, Ricoeur argues, because 'the rejuvenation of all dead metaphors and the invention of new living metaphors that redescribe metaphor allow a new conceptual production to be grafted onto the metaphorical production itself' (1977: 294). This emphasis on living metaphor as the enabler of conceptual production is crucial for Ricoeur in that it focuses attention on the interpretative act: 'Metaphor is living by virtue of the fact that it introduces the spark of imagination into a "thinking more" at the conceptual level. This struggle to "think more", guided by the "vivifying principle", is the "soul" of interpretation' (303). While there is not the space here to fully address Ricoeur's theory of metaphor, it would appear from the discussion of metaphor in this book—especially the 'coining' of the community metaphor—that the inability of Rheingold and others to delimit the meanings and interpretations that accrete around this 'new' metaphor, and the multiple significations it has generated, to some extent complicates Ricoeur's claims. In essence, new metaphor can continue to 'orchestrate' discourse. See Derrida (1998) for a response to Ricoeur's conception of metaphor.

6. At an earlier point in the same text, Derrida adds: '*Spacing* designates *nothing*, nothing that is, no presence at a distance; it is the index of an irreducible exterior, and at the same time of a *movement*, a displacement that indicates an irreducible alterity. I do not see how one could dissociate the two concepts of spacing and alterity' (1981: 81). In a commentary on this concept, Wigley explains: 'Spacing is precisely not space but what Derrida describes as the "becoming space" of that which is meant to be without space (presence, speech, spirit, ideas, and so on). It is that which opens up a space, both in the sense of fissuring an established structure, dividing it or complicating its limits,

but also in the sense of producing space itself as an opening in the tradition. Spacing is at once splintering and productive' (Wigley, 2002: 73). For more critical discussion of the implications of this idea for broader understandings of and engagement with space, see Doreen Massey (2005: 49–54).

7. Henri Lefebvre (2000: 99) makes a similar point: 'Metaphor and metonymy, then. These familiar concepts are borrowed, of course, from linguistics. Inasmuch, however, as we are concerned not with words but rather with space and spatial practice, such conceptual borrowing has to be underwritten by a careful examination of the relationship between space and language.'

8. What concerns Smith and Katz is that most attempts to account for 'multiple locations' are thwarted or undermined insofar as they pose no challenge to underlying conceptions of 'absolute space'. That is to say, they leave 'existing locations rigidly in place', and 'there is no glimpse [. . .] of the combined rupture of received social/geographical space' (1996: 78). Some possible responses to this criticism will be canvassed in the final chapter.

NOTES TO CHAPTER 5

1. As architectural critic Mark Burry writes in 2001, 'The term "cyberspace" [. . .] has today almost reached the level of common language, if not common acceptance for its place as a legitimate architectural construct. [. . .] "Cyber", taken here to mean "computer processed", conjoins the suffix "space", and in doing so propagates the idea of digitally represented realms, at once both realistic and paradoxically elusive. To the observer, these realms may be perceived as tangible (real) or exotically intangible (virtual)' (Burry, 2001: 7).

2. In his contribution to a short edited collection of essays in memory of John Cage, Marcos Novak writes: 'We stand at the dawn of an era that will see the emancipation of architecture from matter. The intuition that allows us to even consider architecture as "frozen music" or music as "molten architecture" comes from a deep and ancient understanding that, in its very essence, architecture exceeds building, as music exceeds sound. Music, especially computer music, will have much to teach the new liquid and gravity-free architecture' (Novak, 1994: 64–65).

3. Illustrative of such a slip is the *Architectural Review*'s coining of the term 'Romantic Pragmatism' to describe certain developments in British architecture in the early 1980s, and which was intended as an oppositional term to the increasingly unfashionable 'Post-Modernism'. In fact, Capon (1999b: 142) remarks on the close similarities in the definitions of the two terms, revealing that in presuming the 'problem' solved little has in fact changed.

4. Other critics offer somewhat different interpretations. For example, Glenn Grant (1990) suggests that transcendence does occur in the novel, albeit via an unlikely route: according to the Situationist notion of 'detournement'. Meanwhile, Scott McQuire (2002: 174) argues that Case, the main protagonist of the novel, 'undergoes an extreme experience paralleling the traditional quest, in which the (male) hero experiences dissolution of ego as the pre-condition for spiritual rebirth' (see also, Voller, 1993; Olsen, 1991: 284).

NOTES TO CHAPTER 6

1. The notion of a 'moral economy of the household' is one to which discussion shall return later in this chapter. In contrast, see Douglas (1991) on why a household is not a monetary economy and cannot use market reasoning.

2. The notion of 'home' as 'utopian community' has some resonance with Wright's (1991: 219–220) description of 'home' as an 'ideal'.

3. For example, writing in the early 1960s, architects Serge Chermayeff and Christopher Alexander argue that domestic privacy is fundamental to wider social health. 'It is [. . .] only through the restored opportunity for firsthand experience that privacy gives can health and sanity be brought back to the world of mass culture. Privacy is most urgently needed and most critical in the place where people live, be it house, apartment, or any other dwelling.' (Chermayeff and Alexander, 1963: 37)

4. Others have put this slightly differently, emphasizing the ambivalences of home: the 'home is neither a natural and desirable setting for women nor the locus that lays the foundation for their alienation but also a place of creation and of recreation of women; a fortress, prison, a place of conflict and of security' (Muñoz González, 2005).

5. The emerging tension between public and private is evident in the designs of many of the stately homes of England during the eighteenth and nineteenth centuries. Lawrence Stone (1991) gives a solid account of these schemes and the 'clear conflict between two simultaneously held ideals: that of the house as a museum for display, in order to enhance prestige, and therefore open to the educated and genteel public; and that of the house as a private home, reserved for the family and their guests' (249). In respect to the latter, Stone notes the use of 'new technology' to achieve privacy, such as 'bell wires, plumbing, lighting, and central heating' (251).

6. Werner (1987) argues this process of internal spatial rearrangement of the domestic home for the purposes of increased privacy has impacted significantly and in manifold ways on familial interpersonal relationships.

7. This includes entry into the home of early forms of electric communications, as well as pre-digital communications technologies such as letters, calling cards, and postcards. On the former, see Marvin (1988) and de Sola Pool (1983). Earlier epistolary technologies of communication are examined in detail in Milne (2010).

8. As Lawrence (1987: 163) notes: 'Both spaces *and objects* can express private/personal and public/shared meanings and values, because the home is simultaneously a haven for withdrawal from society and a credential for esteem and the respect of others [. . .]. From this perspective, the notion of privacy can be interpreted not only in terms of the dialectical relations between spaces and activities inside and outside homes but also in terms of individual and communality and public and private dimensions'.

9. Smart house technologies received a great deal of press in the mid-to-late 1990s. For a representative sample of such reportage in the Australian press, see Keenan (1999), Barker (1999), Plunkett (1994), Eccleston (2000), and 'This is the Toaster' (1995).

10. The work of the American designer/architects Diller and Scofidio presents a strident critique of such projected outcomes. As David Morley explains: 'Diller and Scofidio militantly reject the conventional Taylorist injunction to "eliminate inefficiency . . . in all our daily acts" and to thus "achieve expediency by eliminating all repetition and redundancy". Rather, they are interested in exploring deliberately "ineffecient technologies", or "technologies which produce nothing"—except, most importantly, a strategically reframed and heightened sense of the everyday conventions we take for granted' (Morley, 2007: 254; see also, Betsky, Hays, and Anderson, 2003).

11. Such as McQuire's argument that 'a fault-line stretches between the recurrent desire for home as a stable site, a secure space of shelter and enclosure, and the constant drift towards the frontier as a liminal space of perpetual

transformation and potential conquest. Modern identity belongs neither entirely in the home nor at the frontier, but is split by the psychic and social contradictions of its attachment to these two poles' (McQuire, 1997a: 686).

12. See Pawley (1990: 140–161), and for more broader and more detailed discussion of the phenomenon of technology transfer, see Ihde (1990: 125–139).

13. Such as, to cite one example from among many possible examples, Lynn Spigel's, *Make Room for TV* (1992). This is due in part, Leslie Haddon (1994) argues, to the adherence within much computing research to the same key argument raised about television, that 'it is important to understand the family context of consumption because television is an essentially domestic medium' (86). Haddon proposes that caution be exercised in 'using this starting point when considering other ICTs' (86), and that 'complementary research is required [in order to account for] the popularity, patterns of usage, the meaning and the gendered nature of the home computer [which] arise in large part from processes outside the home' (86). 'So-called "home-computing"', Haddon writes, 'cannot be viewed as an activity based solely in the home' (1994: 86; see also, Kjaer, Halskov Madsen, and Graves Petersen, 2000).

14. As well as impacting on identity formation. This is a point made by Lally (2002) throughout her study of computers in the home, and even more explicitly by other critics. For example, Ian Woodward (2001) argues that 'household objects are interpreted as material elements imbricated in the presentation of a socially plausible and internally consistent aesthetic self'.

15. Just as earlier researchers have done before her. For example, Murdock, Harmann, and Gray describe their own study of home computing as affording 'fleeting glimpses of complex processes, deeply embedded in the sedimented structures of families' interior lives' (1994: 158).

16. This is a process with important gender implications. For instance, in commenting on the links between the domestic and the formal economies, Jane Wheelock notes: 'The process by which judgements about values with regard to personal computers are made is thus part of a complex interaction between formal and complementary economic institutions. The outcome is in part a reproduction of traditional gender and generational distribution patterns in the two sectors, in part a modification of them' (Wheelock, 1994: 111).

17. This 'permeability', it has been argued, is furthered by the role of teletechnologies in the increased blurring of boundaries between home and work (Ellison, 2004; Nippert-Eng, 1996).

18. Of course, this is also true of the television when it was first introduced into the domestic home (see Spigel, 1990, 2002).

19. For further analysis and discussion of these processes, see Arnold (2004) and Nansen, et al. (2009).

20. This concern for both medium and message—the technologies themselves and the uses to which these technologies are put—is, they argue, the point of difference which distinguishes the domestication model from other, broadly 'technological determinist' understandings of how new media rework or 'remediate' old media (Silverstone and Haddon, 1996: 62).

21. In support of this argument, Nicola Green argues that in temporal terms, while mobile technologies 'offer new ways of acting in and perceiving time and space, the practical construction of mobile time in everyday life remains firmly connected to well-established time-based social practices, whether these be institutional (such as clock time, "work time") or subjective (such as "family time")' (Green, 2002: 281).

22. In *Place and Placelessness*, Relph (1986: 4) describes place as often understood as a 'naïve and variable expression of geographical experience'.

23. As one critic puts it, 'cellphones and their connectivity in the world at large are the first high-tech acknowledgment of realspace in the age of cyberspace [with "realspace" synonymous in this context with geography and place in the sense that Doreen Massey means it]. Where the choice was once communication, indoors, away from the physical world, or movement and transportation out in the world with no communication, cellphones open up a third possibility—the world outdoors with full communication' (Levinson, 2003: 5).

24. In making this point, Morley draws on Yi-Fu Tuan's distinction 'between "conversations" (substantive talk about events and issues: a discourse of the public realm) and "chatter" (the exchange of gossip principally designed to maintain solidarity between those involved in the exchange: what Tuan calls a "discourse of the hearth")' (Morley, 2003: 452; see also, Humphreys, 2005).

25. See, for example, Ito (2003) and Yoon (2003). To cite a filmic example, these findings bring to mind Amy Heckerling's *Clueless* (1995), where Cher (Alicia Silverstone) and her friends communicate via mobile phone in the same high school corridor space.

26. See also Lim and Chung (2004), who suggest that, in the case of Singapore at least, this 'softening of time' can have broader cultural implications. They argue that the 'polychronic' nature of mobile phone use unsettles an otherwise 'relatively strict adherence to linear time' in Singaporean culture according to which time is 'considered tangible and is constantly budgeted by individuals' (37).

27. Two examples support this observation. The first is research on video game playing at a Local Area Network (LAN) event in the Netherlands which discovers that those attending the event are motivated to do so in part by a desire for face-to-face social contact (Jansz and Martens, 2005). The second example is drawn from an essay by technology writer Katie Hafner who, reflecting back on her long-term involvement in the WELL virtual community, writes that 'the two are not interchangeable, that physical communities are very much alive and their importance endures. People still want a sense of place, a sense of belonging, in a physical way' (Hafner, 2004).

28. In the mid-1950s, Norwegian architect Christian Norberg-Schulz also experimented with 'flexibility of plan and adaptability of use' within the domestic home. In 1955 he and Ärne Korsmö designed three houses near Oslo which incorporated internal 'flexibility of use' in their design to accommodate 'the day-to-day mobility of the elements of the house and its equipment'. This included provision for media technologies of the day, such as the projection of 'slides, films, etc.' (see P. R. B., 1957: 84 & 130)

29. In addition to televisual considerations, there is also an argument for returning to the history of experimental urban critique and exploration—especially as practised by the likes of Fluxus and the Situationists—in order to better understand what is at stake phenomenologically in contemporary, networked mobility.

30. In many respects such claims dovetail with Augé's (1995) belief that, increasingly, our experience (especially through mobility) of the contemporary urban environment is characterised by an engagement with anonymous and soulless 'non-places'.

NOTES TO CHAPTER 7

1. For Jean-Francois Lyotard, as Andrew Murphie and John Potts (2003: 209) explain, 'Computers could [. . .] provide the information necessary for individuals and groups to decide on the best tactics for "imaginative invention"'.

As Murphie and Potts remark, in holding to this belief, 'Lyotard looks to a creative future rather than a mourning of the past. He calls to us to use technology not to become more utilitarian or profitable, but rather to become more sophisticated, imaginative and inventive in the way we live' (209).

2. What I am proposing here has some resonance with Scott Lash's notion of a 'groundless ground', which he develops in his book *Another Modernity, A Different Rationality* (1999). In this book, Lash develops this notion of a 'groundless ground' within the frame of what he terms a 'second modernity' that 'is not anti-rational or rational but has a principle of rationality based on reflexivity' (3). Important to this 'second modernity', he argues, is the 'dimension of *the ground*' (5—original emphasis). He writes: 'This second, grounded dimension of the other modernity is indeed modern, and not traditional. Only in modernity, and indeed after one or two centuries of high modernity, was it possible to achieve the sort of distance on tradition, on community, on place, to allow it to enter meaningfully into discourse. This ground—which alternately takes the form of community, history, tradition, the symbolic, place, the material, language, life-world, the gift, Sittlichkeit, the political, the religious, forms of life, memory, nature, the monument, the path, fecundity, the tale, habitus, the body—is just as important a dimension of the second modernity as groundlessness. It has, however, been too much forgotten by cultural theory and reflexive sociology' (5–6). Lash's project is to 'retrieve this ground' (6): 'The other modernity, with its different rationality, is a question of the groundless ground. It is a simultaneous movement of deconstruction and retrieval. It retrieves and deconstructs at the same time. It is also a grounded space of a certain consistency. Modernity's fate is eternally to retrieve and eternally to deconstruct the ground' (6). While Lash's book and this one differ in many key respects (both philosophical and otherwise), they do perhaps share a similar spirit and a similar commitment to this notion of a 'groundless ground' of 'deconstruction and retrieval'.

3. Or, as Haber puts it several pages later, a further way of 'excluding coherent subjects is to exclude the possibility of community' (1994: 120). For Nancy, as Ian James (2006: 177) explains, this is because 'subjectivity, space, and community are intimately bound up with each other, that is, each mutually implies or codetermines the other'. Nancy's concern, James argues, is 'that of reworking familiar concepts in relation to a thinking of finitude and within a displacement or questioning of their traditional metaphysical underpinnings' (177).

4. Malpas's philosophical reflections on place are developed in even greater detail in *Heidegger's Topology* (2006), a book he views as a companion volume to his earlier *Place and Experience* (1999).

5. Although, in other respects, Massey is critical of Derrida's understanding of space and 'spacing' (2005: 49–54).

Bibliography

Aarseth, E. (1997) *Cybertext: Perspectives on Ergodic Literature*, Baltimore, MD: The Johns Hopkins University Press.

Abrams, M. H. (1968) 'English Romanticism: The Spirit of the Age', in N. Frye (ed.) *Romanticism Reconsidered: Selected Papers from the English Institute*, 4th reprint, New York: Columbia University Press, pp. 26–72.

'A Building That Moves in the Night' (1981), *New Scientist*, 89(1245), 19 March: 743.

Ackerman, J. S. (1990) *The Villa: Form and Ideology of Country Houses*, Princeton, NJ: Princeton Architectural Press.

Adams, P. (1996) 'Virtual Geography', *21C*, 1: 34–37.

Adams, P. C. (1997) 'Cyberspace and Virtual Places', *The Geographical Review*, 87(2), April: 155–171.

———. (1998) 'Network Topologies and Virtual Place', *Annals of the Association of American Geographers*, 88(1): 88–106.

Adams, P. C., and Warf, B. (1997) 'Introduction: Cyberspace and Geographical Space', *The Geographical Review*, 87(2), April: 139–145.

Agamben, G. (1993) *The Coming Community*, trans. M. Hardt, Minneapolis, MN: University of Minnesota Press.

Agre, P. E. (1998) 'Designing Genres for New Media: Social, Economic, and Political Contexts', in S. G. Jones (ed.), *Cybersociety 2.0: Revisiting Computer-Mediated Communication and Community*, Thousand Oaks: Sage, pp. 69–99.

Ahmed, S., and Fortier, A.-M. (2003) 'Re-imagining Communities', *International Journal of Cultural Studies*, 6(3): 251–259.

Alexander, C. (1965) 'The Question of Computers in Design', *Landscape*, 14(3), Spring: 6–8.

Allford, D. (1984) 'The Creative Iconoclast', in C. Price, *Cedric Price: Works II*, London: Architectural Association, p. 7.

Allon, F. (2001) 'An Ontology of Everyday Control: Living and Working in the "Smart House"', *Southern Review*, 34(3): 8–21.

Althusser, L., and Balibar, E. (1970) *Reading Capital*, trans. B. Brewster, London: NLB.

Altman, R. (1989) 'The Problem of Genre History', in R. Altman (ed.), *The American Hollywood Musical*, Bloomington, IN: Indiana University Press, pp. 90–128.

———. (2004) *Silent Film Sound*, New York: Columbia University Press.

Anderson, B. (1983) *Imagined Communities: Reflections on the Rise and Spread of Nationalism*, London: Verso.

Anderson, K., and Smith, S. J. (2001) 'Editorial: Emotional Geographies', *Transactions of the Institute of British Geographers*, 26: 7–10.

Andrejevic, M. (2005) 'Nothing Comes Between Me and My CPU: Smart Clothes and "Ubiquitous" Computing', *Theory, Culture and Society*, 22(3): 101–119.

Antoniades, A. C. (1992) *Poetics of Architecture: Theory of Design*, New York: Van Nostrand Reinhold.

Appadurai, A. (1990) 'Disjunction and Difference in the Global Cultural Economy', *Public Culture*, 2(2): 1–24.

Arakawa, and Gins, M. (1998) 'The Architectural Body in a Reversible Destiny City', *Architectural Design* (Architectural Design Profile No. 136: 'Architects in Cyberspace II'), 68(11/12), November-December: 43–44.

Arendt, H. (1973) *The Human Condition*, 8th impression, Chicago: The University of Chicago Press.

Argyle, K. (1996) 'Life after Death', in R. Shields (ed.), *Cultures of Internet: Virtual Spaces, Real Histories, Living Bodies*, London: Sage, pp. 133–142.

Argyle, K., and Shields, R. (1996) 'Is There a Body in the Net?', in R. Shields (ed.), *Cultures of Internet: Virtual Spaces, Real Histories, Living Bodies*, London: Sage, pp. 58–69.

Armstrong, R. (1999) 'Science-Fiction Architecture', *Architectural Design* (Architectural Design Profile No. 138: 'Sci-Fi Architecture'), 69(3/4), March–April: 20–21.

Arnold, M. (2004) 'The Connected Homes Project: Probing the Effects and Affects of Domesticated ICTs', in A. Bond (ed.) *Artful Integration: Interweaving Media, Materials and Practices, Vol. 2*. Proceedings of the Eighth Biennial Participatory Design Conference, Toronto.

Ascott, R. (1995) 'The Architecture of Cyberception', *Architectural Design* (Architectural Design Profile No. 118: 'Architects in Cyberspace'), 65(11/12), November–December: 38–41.

———. (1998) 'Technoetic Structures', *Architectural Design* (Architectural Design Profile No. 136: 'Architects in Cyberspace II'), 68(11/12), November-December: 30–32.

Augé, M. (1995) *Non-Places: Introduction to an Anthropology of Supermodernity*, trans. J. Howe, London: Verso.

Aurigi, A. (2005) *Making the Digital City: The Early Shaping of Urban Internet Space*, Aldershot, Hampshire, England: Ashgate.

Bakardjieva, M., and Smith, R. (2001) 'The Internet in Everyday Life: Computer Networking from the Standpoint of the Dometic User', *New Media and Society*, 3(1): 67–83.

Bailey, C. (1996) 'Virtual Skin: Articulating Race in Cyberspace', in M. A. Moser and D. MacLeod (eds), *Immersed in Technology: Art and Virtual Environments*, Cambridge, MA: MIT Press, pp. 29–49.

Baillieu, A. (1998) 'Bye Bye Baby', *RIBA Journal*, 105(2): 6–9, 11.

Bammer, A. (1992) 'Editorial', *New Formations*, 17, Summer: vii–xi.

Barker, G. (1999) 'The Techno-Servants Have Arrived—With More to Come', *Sunday Age*: Property, 17 October: 3.

———. (2006) 'The Mobile Phone Revolution', *Age*: LiveWire, 25 May, <http://www.theage.com.au/news/phones—pdas/mobile-revolution/2006/05/24/1148150262957.html> (accessed 27 November 2010).

Barlow, J. P. (1999) 'Is There a There in Cyberspace?', *Utne*, 15 September, <http://www.utne.com/archives/IsThereaThereinCyberspace.aspx> (accessed 27 November 2010).

Barthes, R. (1986) *The Rustle of Language*, trans. R. Howard, Oxford: Basil Blackwell.

Bauman, Z. (2001) *Community: Seeking Safety in an Insecure World*, Oxford: Polity.

Baym, N. (1998) 'The Emergence of On-Line Community', in S. G. Jones (ed.), *Cybersociety 2.0: Revisiting Computer-Mediated Communication and Community*, Thousand Oaks: Sage, pp. 35–68.

———. (2002) 'Interpersonal Life Online', in L. A. Lievrouw and S. Livingstone (eds), *Handbook of New Media: Social Shaping and Consequences of ICTs*, London: Sage, pp. 62–76.

Beaubien, M. P. (1996) 'Playing at Community: Multi-User Dungeons and Social Interaction in Cyberspace', in L. Strate, R. Jacobson, and S. B. Gibson (eds), *Communication and Cyberspace: Social Interaction in an Electronic Environment*, Cresskill, NJ: Hampton Press, pp. 179–188.

Bell, C., and Newby, H. (1975) *Community Studies: An Introduction to the Sociology of the Local Community*, 2nd impression, London: George Allen & Unwin Ltd.

Bell, D. (2001) *An Introduction to Cybercultures*, London: Routledge.

Bell, J. (1996) 'Architecture of the Virtual Community', Centre for Built Environment, Cardiff University, Cardiff, Wales, <http://ctiweb.cf.ac.uk/dissertations/virtual_architecture/contents.html> (accessed 20 June 2001).

Benedikt, M. (1992a) 'Cyberspace: Some Proposals', in M. Benedikt (ed.), *Cyberspace: First Steps*, Cambridge, MA: MIT Press, pp. 119–224.

———. (1992b) 'Introduction', in M. Benedikt (ed.), *Cyberspace: First Steps*, Cambridge, MA: The MIT Press, pp. 1–25.

———. (ed.) (1992c) *Cyberspace: First Steps*, Cambridge, MA: MIT Press.

Benford, S., Bowers, J., Fahlén, L. E., Greenhalgh, C., and Snowdon, D. (1995) 'User Embodiment in Collaborative Virtual Environments', in I. R. Katz, R. Mack, L. Marks, M. B. Rosson, and J. Nielsen (eds), *Human Factors in Computing Systems: CHI '95 Conference Proceedings*, New York: The Association for Computing Machinery (ACM), pp. 242–249.

Bentley, N. (1999) 'A Virtual Community', *Melbourne University Magazine*, Spring: 20.

Bernard, J. (1973) *The Sociology of Community*, Glenview, IL: Scott, Foresman and Company.

Berry, D. (1995) 'The Poetics of Cyberspace (Draft)', CyberUrbanity list archive, 28 April, <http://platon.ee.duth.gr/data/maillist-archives/cyberurbanity/1995/msg00109.html> (accessed 2 October 2000).

Betsky, A. (1999) 'Freed Form', *Wired*, September: 168–173.

Betsky, A., Hays, K. M., Anderson, L. (eds) (2003) *Scanning: The Aberrant Architectures of diller+scofidio*, New York: Whitney Museum of American Art.

Blackburn, S. (1996) *The Oxford Dictionary of Philosophy*, Oxford: Oxford University Press.

Blanchot, M. (1987) 'Everyday Speech', trans. S. Hanson, *Yale French Studies*, 73: 12–20.

———. (1988) *The Unavowable Community*, trans. P. Joris, Barrytown, New York: Station Hill Press.

Blavin, J. H., and Cohen, I. G. (2002) 'Gore, Gibson, and Goldsmith: The Evolution of Internet Metaphors in Law and Commentary', *Harvard Journal of Law and Technology*, 16(1), Fall: 265–285.

Boden, D., and Molotch, H. L. (1994) 'The Compulsion of Proximity', in R. Friedland and D. Boden (eds), *NowHere: Space, Time and Modernity* (Berkeley, California: University of California Press, pp. 257–286.

Bollier, D. (1995) *The Future of Community and Personal Identity in the Coming Electronic Culture: A Report of the Third Annual Aspen Institute Roundtable on Information Technology, Aspen, Colorado, August 18–21, 1994*, Washington, DC: The Aspen Institute.

Bolter, J. D., and Grusin, R. (1999) *Remediation: Understanding New Media*, Cambridge, MA: MIT Press.

Bowra, M. (1963) *The Romantic Imagination*, London: Oxford University Press.

Boyarin, D., and Boyarin, J. (1993) 'Diaspora: Generation and the Grounds of Jewish Identity', *Critical Inquiry*, 19, Summer: 693–725.

Boyer, M. C. (1996) *Cybercities: Visual Perception in the Age of Electronic Communications*, New York: Princeton University Press.

Brabazon, T. (2001) 'How Imagined Are Virtual Communities?' *Mots Pluriels*, 18, August, <http://www.arts.uwa.edu.au/MotsPluriels/MP1801tb2.html> (accessed 27 November 2010).

Bringing Down a Dictator (2001) film, directed by S. York, Washington, DC: York Zimmeman Inc.

Brodsky Lacour, C. (1996) *Lines of Thought: Discourse, Architectonics and the Origins of Modern Philosophy*, Durham, NC: Duke University Press.

———. (1999) 'Architecture in the Discourse of Modern Philosophy: Descartes to Nietzsche', in A. Kosta and I. Wohlfarth (eds), *Nietzsche and 'An Architecture of Our Minds'*, Los Angeles: The Getty Research Institute for the History of Art and the Humanities, pp. 19–34.

Bromberg, H. (1996) 'Are MUDs Communities?: Identity, Belonging and Consciousness in Virtual Worlds', in R. Shields (ed.), *Cultures of Internet: Virtual Spaces, Real Histories, Living Bodies*, London: Sage, pp. 143–152.

Bruckman, A. (1996) 'Finding One's Own Space in Cyberspace', *Technology Review*, 99(1): 48–54.

Bryson, L., and Mowbray, M. (1981) '"Community": The Spray-on Solution', *Australian Journal of Social Issues*, 16(4): 255–267.

Bull, M. (2001) 'The World According to Sound: Investigating the World of Walkman Users', *New Media and Society*, 3(2): 179–197.

———. (2004) 'Sound Connections: An Aural Epistemology of Proximity and Distance in Urban Culture', *Environment and Planning D: Society and Space*, 22: 103–116.

Burkhalter, B. (1999) 'Reading Race Online', in M. A. Smith and P. Kollock (eds), *Communities in Cyberspace*, London: Routledge, pp. 60–75.

Burnett, R. (1999) 'Communities in Cyberspace: Towards a New Research Agenda', *Continuum*, 13(2), July: 205–216.

Burry, M. (2001) *Cyberspace: The World of Digital Architecture*, Mulgrave, Victoria, Australia: The Images Publishing Group.

Cahnman, W. J. (1995) *Weber and Toennies: Comparative Sociology in Historical Perspective*, J. B. Maier, J. Marcus, and Z. Tarr (eds), New Brunswick: Transaction Publishers.

Calcutt, A. (1999) *White Noise: An A-Z of the Contradictions in Cyberculture*, London: Macmillan.

Campion, D. (1968) *Computers in Architectural Design*, London: Elsevier Publishing Company.

Capon, D. S. (1999a) *Architectural Theory Volume One: The Vitruvian Fallacy*, Chichester, West Sussex: John Wiley & Sons.

———. (1999b) *Architectural Theory Volume Two: Le Corbusier's Legacy*, Chichester, West Sussex: John Wiley & Sons.

Caputo, J. D. (1997) 'Community Without Community', in J. D. Caputo (ed), *Deconstruction in a Nutshell: A Conversation with Jacques Derrida*, New York: Fordham University Press, pp. 106–124.

Carter, D. (2005) 'Living in Virtual Communities: An Ethnography of Human Relationships in Cyberspace', *Information, Communication and Society*, 8(2), June: 148–167.

Casey, E. S. (1993) *Getting Back into Place: Towards a Renewed Understanding of the Place-World*, Bloomington, IN: Indiana University Press.

———. (1997) *The Fate of Place: A Philosophical History*, Berkeley, CA: University of California Press.

Castells, M. (1996) *The Information Age: Economy, Society and Culture: Volume 1: The Rise of the Networked Society*, Oxford: Blackwell.

Castells, M., Fernández-Ardèvol, M., Qiu, J. L., and Sey, A. (2007) *Mobile Communication and Society: A Global Perspective*, Cambridge, MA: MIT Press.

'Cedric Price' (2003), *The Telegraph*, <http://www.telegraph.co.uk/news/obituaries/1438827/Cedric-Price.html> (accessed 27 November 2010).

Chabrán, R., and Salinas, R. (2004) 'Place Matters: Journeys through Global and Local Spaces', in M. Sturken, D. Thomas, and S. J. Ball-Rokeach (eds), *Technological Visions: The Hopes and Fears that Shape New Technologies*, Philadelphia, PA: Temple University Press, pp. 305–338.

Chapman, T. (2001) 'There's No Place Like Home', *Theory, Culture & Society*, 18(6): 135–146.

Chermayeff, S., and Alexander, C. (1963) *Community and Privacy: Toward a New Architecture of Humanism*, Garden City, NY: Anchor Books.

Chermayeff, S., and Tzonis, A. (1971) *Shape of Community: Realization of Human Potential*, Harmondsworth, Middlesex: Penguin.

Cherny, L. (1999) *Conversation and Community: Chat in a Virtual World*, Stanford, CA: Center for the Study of Language and Information.

Chun, W. H. K., and Keenan, T. (eds) (2006) *New Media, Old Media: A History and Theory Reader*, New York: Routledge.

Clark, G. L. (2005) 'Money Flows Like Mercury: The Geography of Global Finance', *Geografiska Annaler: Series B, Human Geography*, 87(2), June: 97–164.

Clueless (1995) film, directed by A. Heckerling, Los Angeles, CA: Paramount Pictures.

Coffman, T. (2003) *Building for Hearst and Morgan: Voices from the George Loorz Papers*, Berkeley, CA: Berkeley Hill Books.

Conrads, U., and Sperlich, H. G. (1963) *Fantastic Architecture*, trans. C. Crasemann Collins and G. R. Collins, London: The Architectural Press.

Cook, P. (1967) *Architecture: Action and Plan*, London: Studio Vista.

———. (1970) *Experimental Architecture*, London: Studio Vista.

Cooley, H. R. (2004) 'It's All About the *Fit*: The Hand, the Mobile Screenic Device and Tactile Vision', *Journal of Visual Culture*, 3(2): 133–155.

Cooper, D. (1996) 'Paper Cities', *21C*, 3: 12–14.

Cooper, S. (2004) 'Cooper's Last: Post-Derrida', *Arena Magazine*, 74, December–January: 33–34.

Corlett, W. (1993) *Community Without Unity: A Politics of Derridean Extravagance*, Durham, NC: Duke University Press.

Couldry, N., and McCarthy, A. (2004) 'Introduction. Orientations: Mapping Mediaspace', in N. Couldry and A. McCarthy (eds), *MediaSpace: Place, Scale and Culture in a Media Age*, London: Routledge, pp. 1–18.

Cowling, D. (1998) *Building the Text: Architecture as Metaphor in Late Medieval and Early Modern France*, Oxford: Clarendon Press.

Coyne, R. (1995) *Designing Information Technology in the Postmodern Age: From Method to Metaphor*, Cambridge, MA: MIT Press.

Crane, D., Kawashima, N., and Kawasaki, K. (eds) (2002) *Global Culture: Media, Arts, Policy, and Globalization*, London: Routledge.

Cresswell, T. (2004) *Place: A Short Introduction*, Oxford: Blackwell.

Cross, N. (1977) *The Automated Architect*, London: Pion.

Crowther, P. (1995) 'The Postmodern Sublime: Installation and Assemblage Art', *Art & Design* (Art and Design Profile No. 40: 'The Contemporary Sublime'), 10(1/2), January–February: 8–17.

Csicsery-Ronay, I. (1988) 'Cyberpunk and Neuromanticism', *Mississippi Review*, 16(2–3): 266–278.

Damer, B. (1998) *Avatars! Designing and Building Virtual Worlds on the Internet*, Berkeley, CA: Peachpit Press.

Damrau, K. (1999) 'Beyond Solidity: Inventions, Spaces and Concepts for the Elements of Air and Water', *Architectural Design* (Architectural Design Profile No. 138: 'Sci-Fi Architecture'), 69(3/4), March–April: 22–33.

Davidson, J., and Milligan, C. (2004) 'Editorial: Embodying Emotion Sensing Space: Introducing Emotional Geographies', *Social and Cultural Geography*, 5(4), December: 523–532.

Davidson, M. (1983) *Uncommon Sense: The Life and Thought of Ludwig Von Bertalanffy (1901–1972), Father of General Systems Theory*, Los Angeles: J. P. Tarcher, Inc.

Davis, M. (1991) *City of Quartz: Excavating the Future in Los Angeles*, London: Verso.

———. (2006) *Planet of Slums*, New York: Verso.

Day, P., and Schuler, D. (eds) (2004) *Community Practice in the Network Society: Local Action / Global Interaction*, London: Routledge.

De Sola Pool, I. (1983) *Forecasting the Telephone: A Retrospective Technology Assessment of the Telephone*, Norwood, NJ: Ablex.

Delanty, G. (2003) *Community*, London: Routledge.

Deleuze, G. (1992) 'Postscript on the Societies of Control', *October*, 59, Winter: 3–7.

———. (1994) *Difference and Repetition*, trans. P. Patton, New York: Columbia University Press.

———. (2002) 'The Actual and the Virtual', trans. P. Patton, in G. Deleuze and C. Parnet, *Dialogues II*, trans. H. Tomlinson and B. Habberjam, New York: Columbia University Press, pp. 148–152.

Derrida, J. (1981) 'Positions', in *Positions*, trans. A. Bass, Chicago: University of Chicago Press, pp. 37–96.

———. (1982a) 'Différance', in *Margins of Philosophy*, trans. A. Bass, Hemel Hempstead, Hertfordshire: Harvester Wheatsheaf, pp. 1–27.

———. (1982b) 'White Mythology: Metaphor in the Text of Philosophy', in *Margins of Philosophy*, trans. A. Bass, Hemel Hempstead, Hertfordshire: Harvester Wheatsheaf, pp. 207–271.

———. (1985) 'Roundtable on Autobiography', in *The Ear of the Other: Otobiography, Transference, Translation*, trans. P. Kamuf, New York: Schocken Books, pp. 39–89.

———. (1987) 'Women in the Beehive: A Seminar with Jacques Derrida', trans. J. Adner, in A. Jardine and P. Smith (eds), *Men in Feminism*, New York: Methuen, pp. 189–203.

———. (1992) *Given Time: I. Counterfeit Money*, trans. P. Kamuf, Chicago: University of Chicago Press.

———. (1994) *Specters of Marx: The State of the Debt, the Work of Mourning, and the New International*, trans. P. Kamuf, New York: Routledge.

———. (1995a) 'Khōra', trans. I. McLeod, in J. Derrida, *On the Name*, T. Dutoit (ed.), Stanford, CA: Stanford University Press, pp. 89–127.

———. (1995b) *The Gift of Death*, trans. D. Wills, Chicago: University of Chicago Press.

———. (1997a) *Of Grammatology*, corrected edition, trans. G. C. Spivak, Baltimore, MD: The Johns Hopkins University Press.

———. (1997b) *Politics of Friendship*, trans. G. Collins, London: Verso.

———. (1998), 'The *Retrait* of Metaphor', trans. F. Gasdner, in J. Wolfreys (ed.), *The Derrida Reader: Writing Performances* (Lincoln, Nebraska: University of Nebraska Press, pp. 102–129.

———. (2001a) 'Force and Signification', in *Writing and Difference*, trans. A. Bass, London: Routledge, pp. 1–35.

———. (2001b) *On Cosmopolitanism and Forgiveness*, trans. M. Dooley and M. Hughes, London: Routledge.

———. (2001c) 'Structure, Sign and Play in the Discourse of the Human Sciences', in *Writing and Difference*, trans. A. Bass, London: Routledge, pp. 351–370.

——. (2005) *Paper Machine*, trans. R. Bowlby, Stanford, CA: Stanford University Press.

Derrida, J., and Caputo, J. D., Schmidt, D., Brogan, W., Busch, T. W., and Murphy, J. (1997) 'The Villanova Roundtable: A Conversation with Jacques Derrida', in J. D. Caputo (ed.), *Deconstruction in a Nutshell: A Conversation with Jacques Derrida*, New York: Fordham University Press, pp. 1–28.

Derrida, J., and Dufourmantelle, A. (2000) *Of Hospitality / Anne Dufourmantelle Invites Jacques Derrida to Respond*, trans. J. Bajorek, Stanford, CA: Stanford University Press.

Derrida, J., and Kipnis, J. (1987) 'Afterword', in J. Kipnis and T. Leeser (eds), *Chora L Works: Jacques Derrida and Peter Eisenman*, New York: The Monacelli Press, pp. 166–172.

Derrida, J., Sohm, B., de Peretti, C., Douailler, S., Vermeren, P., and Malet, E. (1994) 'The Deconstruction of Actuality: An Interview with Jacques Derrida', trans. J. Rée, *Radical Philosophy*, 68, Autumn: 28–41.

Derrida, J., and Stiegler B. (2002) *Echographies of Television: Filmed Interviews*, trans. J. Bajorek, Cambridge, UK: Polity.

Dessauce, M. (ed.) (1999) *The Inflatable Moment: Pneumatics and Protest in '68*, New York: Princeton Architectural Press and The Architectural League of New York.

Devisch, I. (2006) 'The Sense of Being(-)With Jean-Luc Nancy', *Culture Machine*, 8, <http://www.culturemachine.net/index.php/cm/article/view/36/44> (accessed 27 November 2010).

Dewey, R. (1968) 'The Rural-Urban Continuum: Real but Relatively Unimportant', in R. L. Warren (ed.), *Perspectives on the American Community: A Book of Readings*, 4th printing, Chicago: Rand McNally & Company, pp. 184–192.

Doheny-Farina, S. (1996) *The Wired Neighborhood*, New Haven: Yale University Press.

Dollens, D. (2005) *Digital-Botanic Architecture*, Santa Fe, New Mexico: SITES books.

Donini, A. O., and Novack, J. A. (eds) (1982) *Origins and Growth of Sociological Theory: Readings on the History of Sociology*, Chicago, IL: Nelson-Hall.

Douglas, M. (1991) 'The Idea of a Home: A Kind of Space', *Social Research*, 58(1), Spring: 287–307.

Dovey, K. (2002) 'Dialectics of Place: Authenticity, Identity, Difference', in S. Akkach (ed.), *De-Placing Difference: Architecture, Culture and Imaginative Geography*, Adelaide: Centre for Asian and Middle Eastern Architecture, The University of Adelaide, pp. 45–52.

Dyson, E. (1997) *Release 2.1: A Design for Living in the Digital Age*, New York: Broadway Books.

Easthope, H. (2004) 'A Place Called Home', *Housing, Theory and Society*, 21(3): 128–138.

Eccleston, R. (2000) 'Nesting Instincts', *The Australian Magazine*, 1–2 January: 20–22 & 24.

'Editorial: Whatever Happened to the Systems Approach?' (1976), *Architectural Design*, XLVI, May: 267.

Edwards, C. (2005) *Turning Houses into Homes: A History of the Retailing and Consumption of Domestic Furnishings*, Aldershot, Hampshire, England: Ashgate.

Ek, R. (2006) 'Media Studies, Geographical Imaginations and Relational Space', in J. Falkheimer and A. Jansson (eds), *Geographies of Communication: The Spatial Turn in Media Studies*, Göteborg: Nordicom, pp. 45–66.

Eliot, T. S. (1943) *Little Gidding*, London: Faber and Faber.

Ellison, N. B. (2004) *Telework and Social Change: How Technology Is Reshaping the Boundaries between Home and Work*, Westport, CT: Praeger.

Escobar, A. (1996) 'Welcome to Cyberia: Notes on the Anthropology of Cyberculture', in Z. Sardar and J. R. Ravetz (eds), *Cyberfutures: Culture and Politics on the Information Superhighway*, London: Pluto Press: 111–137.

Estévez, A. T., Puigarnau, A., Pérez Arnal, I., Dollens, D., Pérez-Méndez, A., Ruiz Millet, J., and Planella, A. (2003) *Genetic Architectures / Arquitecturas Genéticas*, Santa Fe, New Mexico: SITES books.

Etzioni, A., and Etzioni, O. (1999) 'Face-to-Face and Computer-Mediated Communities, A Comparative Analysis', *The Information Society*, 15: 241–248.

Fair, G. R., Flowerdew, A. D. J., Munro, W. G., and Rowley, D. (1966) 'Note on the Computer as an Aid to the Architect', *The Computer Journal*, 9, May: 16–20.

Fathy, T. A. (1991) *Telecity: Information Technology and Its Impact on City Form*, New York: Praeger.

Feenberg, A., and Bakardjieva, M. (2004) 'Virtual Community: No "Killer Implication"', *New Media and Society*, 6(1): 37–43.

Ferguson, F. (1975) *Architecture, Cities and the Systems Approach*, New York: George Braziller.

Fernback, J. (1999) 'There Is a There There', in S. G. Jones (ed.), *Doing Internet Research: Critical Issues and Methods for Examining the Net*, Thousand Oaks: Sage, pp. 203–220.

Figallo, C. (1998) *Hosting Web Communities: Building Relationships, Increasing Customer Loyalty, and Maintaining a Competitive Edge*, New York: John Wiley and Sons.

Fleckenstein, K. S. (1995) 'Writing and the Strategic Use of Metaphor', *TETYC*, May: 110–115.

Flew, T. (2001) 'The "New Empirics" in Internet Studies', in H. Brown, G. Lovink, H. Merrick, N. Rossiter, D. Teh and M. Willson (eds), *Politics of a Digital Present: An Inventory of Australian Net Culture*, Melbourne: Fibreculture Publications, pp. 105–113.

Flusser, V. (1990) 'Curie's Children: Vilém Flusser on Future Architecture', *Artforum*, XXVIII(9), May: 35–36.

Forty, A. (2004) *Words and Buildings: A Vocabulary of Modern Architecture*, London: Thames and Hudson.

Foth, M. (ed.) (2009) *Handbook of Research on Urban Informatics: The Practice and Promise of the Real-Time City*, Hershey, PA: Information Science Reference, IGI Global.

Foucault, M. (1986) 'Of Other Spaces', trans. J. Miskowiec, *Diacritics: A Review of Contemporary Criticism*, 16(1), Spring: 22–27.

Franck, K. A. (1995) 'When I Enter Virtual Reality, What Body Will I Leave Behind?', *Architectural Design* (Architectural Design Profile No. 118: 'Architects in Cyberspace'), 65(11/12), November–December: 20–23.

———. (1998) 'It and I: Bodies as Objects, Bodies as Subjects', *Architectural Design* (Architectural Design Profile No. 136: 'Architects in Cyberspace II'), 68(11/12), November–December: 16, 18–19.

Frazer, J. (1995) *Themes VII: An Evolutionary Architecture*, London: Architectural Association.

———. (2003) 'The Continuing Relevance of Generator—The Archetypal Generator', in S. Hardingham (ed.), *Cedric Price Opera*, Chichester, West Sussex: Wiley-Academy, pp. 46–48.

Freud, S. (1971) 'The "Uncanny"', trans. A. Strachey, in J. Strachey, A. Freud, A. Strachey, and A. Tyson (eds), *The Standard Edition of the Complete Psychological Works of Sigmund Freud, Volume XVII (1917–1919): An Infantile Neurosis and Other Works*, London: The Hogarth Press, pp. 218–256.

Freund, J. (1978) 'German Sociology in the Time of Max Weber', in T. Bottomore and R. A. Nisbet (eds), *A History of Sociological Analysis*, London: Heinemann, pp. 149–186.

Friedberg, A. (1993) *Window Shopping: Cinema and the Postmodern*, Berkeley, CA: University of California Press.

Friedman, Y. (1975) *Toward a Scientific Architecture*, trans. C. Lang, Cambridge, MA: MIT Press.

Froehling, O. (1997) 'The Cyberspace "War of Ink and Internet" in Chiapas, Mexico', *Geographical Review*, 87(2), April: 291–307.

Frye, N. (1968) 'The Drunken Boat: The Revolutionary Element in Romanticism', in N. Frye (ed.), *Romanticism Reconsidered: Selected Papers from the English Institute*, 4th reprint, New York: Columbia University Press, pp. 1–25.

Fynsk, C. (1991) 'Foreword: Experiences of Finitude', in J.-L. Nancy, *The Inoperative Community*, trans. P. Connor, L. Garbus, M. Holland and S. Sawhney, Minneapolis, MN: University of Minnesota Press, pp. vii–xxxv.

Gage, S. (1998) 'Intelligent Interactive Architecture', *Architectural Design* (Architectural Design Profile No. 136: 'Architects in Cyberspace II'), 68(11/12), November–December: 81–84.

Galloway, A. (2004) 'Intimations of Everyday Life: Ubiquitous Computing and the City', *Cultural Studies*, 18(2/3), March–May: 384–408.

———. (2006) 'Thingers Rather Than Thinkers I', *Purse Lip Square Jaw* [weblog], 16 February, <http://www.purselipsquarejaw.org/2006/02/thingers-rather-than-thinkers-i.php> (accessed 27 November 2010).

Galloway, A., and Ward, M. (2006) 'Locative Media as Socialising and Spatializing Practice: Learning from Archaeology', *Leonardo Electronic Almanac*, 14(3) <http://www.leoalmanac.org/journal/vol_14/lea_v14_n03–04/gallowayward.html> (accessed 27 November 2010).

Gandhi, L. (2003) 'Friendship and Postmodern Utopianism', *Cultural Studies Review*, 9(1): 12–22.

———. (2006) *Affective Communities: Anticolonial Thought, Fin-de-Siècle Radicalism, and the Politics of Friendship*, Durham, NC: Duke University Press.

Gates, B. (1996) *The Road Ahead*, New York: Penguin.

Gibson, W. (1993a) *Neuromancer*, London: HarperCollins.

———. (1993b) *Virtual Light*, London: Viking.

Gillespie, M. (2000) 'Transnational Communications and Diaspora Communities', in S. Cottle (ed.), *Ethnic Minorities and the Media*, Buckingham: Open University Press, pp. 164–178.

Gitelman, L. (2008) *Always Already New: Media, History, and the Data of Culture*, Cambridge, MA: MIT Press.

Gitelman, L., and Pingree, G. B. (eds) (2003) *New Media, 1740–1915*, Cambridge, MA: MIT Press.

Glazer, M. (1989) '"What Is Within Now Seen Without": Romanticism, Neuromanticism, and the Death of the Imagination in William Gibson's Fictive World', *Journal of Popular Culture*, 23(3), Winter: 155–164.

Glowacka, D. (2006) 'Community: *Comme-un?*', *Culture Machine*, 8, <http://www.culturemachine.net/index.php/cm/article/view/34/42> (accessed 27 November 2010).

Goggin, G. (2006) 'SMS Riot: Transmitting Race on a Sydney Beach, December 2005', *M/C Journal*, 9(1), < http://journal.media-culture.org.au/0603/02-goggin.php> (accessed 5 November 2006).

———. (2011) *Global Mobile Media*, New York: Routledge.

Goggin, G., and McLelland, M. (2009a) 'Internationalizing Internet Studies: Beyond Anglophone Paradigms', in G. Goggin and M. McLelland (eds), *Internationalizing Internet Studies*, New York: Routledge, pp. 3–17.

———. (eds) (2009b) *Internationalizing Internet Studies*, New York: Routledge.

Gombrich, E. H. (1972) *Art and Illusion: A Study in the Psychology of Pictorial Representation*, 4th edition, London: Phaidon.

Goode, W. J. (1957) 'Community within Community: The Professions', *American Sociological Review*, 22(2), April: 194–200.

Graham, G. (1999) *The Internet: A Philosophical Inquiry*, London: Routledge.

Graham, S. (2004) 'Beyond the "Dazzling Light": From Dreams of Transcendence to the "Remediation" of Urban Life', *New Media and Society*, 6(1): 16–25.

Grant, G. (1990) 'Transcendence through Detournement in William Gibson's *Neuromancer*', *Science-Fiction Studies*, 17(1), March: 41–49.

Grass, G. (1979) *The Flounder*, Harmondsworth, Middlesex: Penguin.

Green, N. (2002) 'On the Move: Technology, Mobility, and the Mediation of Social Time and Space', *The Information Society*, 18(4): 281–292.

Grossman, W. M. (1997) *Net.Wars*, New York: New York University Press.

Grosz, E. (1998) 'Thinking the New: Of Futures Yet Unthought', *Symploke*, 6(1): 38–55.

Guallart, V. (2004) *Media House Project: The House Is the Computer, the Structure Is the Network*, Barcelona: Institut d'arquitectura avançada de Catalunya.

Gumpert, G., and Drucker, S. J. (1996) 'From Locomotion to Telecommunication, or Paths of Safety, Streets of Gore', in L. Strate, R. Jacobson, and S. B. Gibson (eds), *Communication and Cyberspace: Social Interaction in an Electronic Environment*, Cresskill, NJ: Hampton Press, pp. 25–37.

Gurstein, M. (ed.) (2000) *Community Informatics: Enabling Communities with Information and Communications Technologies*, Hershey, PA: Idea Group Publishing.

Gusfield, J. R. (1975) *Community: A Critical Response*, Oxford: Basil Blackwell.

Gye, L. (2004) '2004: Glorious Corporeality', *RealTime: OnScreen / Digital*, 63, October–November: 33.

Gye, L., and Tofts, D. (2003) 'The Holographic Nursery', *MESH*, 16, <http://www.experimenta.org/hot/melbourne/mesh.html> (accessed 27 November 2010).

Haber, H. F. (1994) *Beyond Postmodern Politics: Lyotard, Rorty, Foucault*, London: Routledge.

Haddon, L. (1994) 'Explaining ICT Consumption: The Case of the Home Computer', in R. Silverstone and E. Hirsch (eds), *Consuming Technologies: Media and Information in Domestic Spaces*, London: Routledge, pp. 82–96.

Hafner, K. (2004) 'When the Virtual Isn't Enough', in M. Sturken, D. Thomas, and S. J. Ball-Rokeach (eds), *Technological Visions: The Hopes and Fears that Shaped New Technologies*, Philadelphia, PA: Temple University Press, pp. 293–304.

Hagel, J., and Armstrong, A. G. (1997) *Net Gain: Expanding Markets through Virtual Communities*, Boston, MA: Harvard Business School Press.

Hampton, K. (2002) 'Place-based and IT Mediated Community', *Planning Theory and Practice*, 3(2): 228–231.

Hampton, K. N., and Wellman, B. (2000) 'Examining Community in the Digital Neighborhood: Early Results from Canada's Wired Suburb', in T. Ishida and K. Isbister (eds), *Digital Cities: Technologies, Experiences, and Future Perspectives*, Heidelberg, Germany: Springer-Verlag, pp. 194–208.

Handler, A. B. (1970) *Systems Approach to Architecture*, New York: American Elsevier Publishing Company.

Haraway, D. J. (1991) *Simians, Cyborgs, and Women: The Reinvention of Nature*, London: Free Association Books.

Harbison, R. (1992) *The Built, the Unbuilt, and the Unbuildable: In Pursuit of Architectural Meaning*, Cambridge, MA: MIT Press.

Hardt, M., and Negri, A. (2000) *Empire*, Cambridge, MA: Harvard University Press.

Hareven, T. K. (1991) 'The Home and the Family in Historical Perspective', *Social Research*, 58(1), Spring: 253–285.

Harrison, T. M., and Stephen, T. (1999) 'Researching and Creating Community Networks', in S. G. Jones (ed.), *Doing Internet Research: Critical Issues and Methods for Examining the Net*, Thousand Oaks: Sage, pp. 221–241.

Harvey, D. (1990) 'Between Space and Time: Reflections on the Geographical Imagination', *Annals of the Association of American Geographers*, 80(3): 418–434.

———. (2000) *Spaces of Hope*, Edinburgh: Edinburgh University Press.

Hauben, M., and Hauben, R. (1997) *Netizens: On the History and Impact of Usenet and the Internet*, Los Alamitos, CA: IEEE Computer Society Press.

Hayles, N. K. (1996) 'Embodied Virtuality', in M. A. Moser and D. MacLeod (eds), *Immersed in Technology: Art and Virtual Environments*, Cambridge, MA: MIT Press, pp. 1–28.

Haythornthwaite, C. (2002) 'Strong, Weak, and Latent Ties and the Impact of New Media', *The Information Society*, 18(5): 385–401.

Heidegger, M. (1971a) *Poetry, Language, Thought*, trans. A. Hofstadter, New York: Harper & Row.

———. (1971b) 'What are Poets For?', in *Poetry, Language, Thought*, trans. Albert Hofstadter, New York: Harper & Row, pp. 89–142.

Heim, M. (1993) *The Metaphysics of Virtual Reality*, New York: Oxford University Press.

Held, D., and McGrew, A. (2002) *Globalization/Anti-Globalization*, Cambridge, UK: Polity.

Hellman, L. (1986) *Architecture for Beginners*, London: Writers & Readers, in association with Unwin Paperbacks.

Hennessey, P. (2000) 'Datatecture: The Metaphorical Architecture of Cyberspace', *Metro*, 121/122: 69, 71–76.

Hetherington, K. (1994) 'The Contemporary Significance of Schmalenbach's Concept of the Bund', *The Sociological Review*, 42(1), February: 1–25.

———. (1998) *Expressions of Identity: Space, Performance, Politics*, London: Sage.

Highmore, B. (2002) *Everyday Life and Cultural Theory: An Introduction*, London: Routledge.

Higuchi, T. (1983) *The Visual and Spatial Structure of Landscapes*, trans. C. S. Terry, Cambridge, MA: MIT Press.

Hillery, G. A. (1955) 'Definitions of Community: Areas of Agreement', *Rural Sociology*, 20(1): 111–123.

Hine, C. (2002) *Virtual Ethnography*, London: Sage.

Hjorth, L. (2005) 'Society of the Phoneur', *antiTHESIS*, 15: 208–215.

Hoem, S. I. (1996) 'Community and the "Absolutely Feminine"', *Diacritics: A Review of Contemporary Criticism*, 26, Summer: 49–58.

Hollier, D. (1989) *Against Architecture: The Writings of Georges Bataille*, trans. B. Wing, Cambridge, MA: MIT Press.

Holmes, D. (1997) 'Introduction: Virtual Politics—Identity and Community in Cyberspace', in D. Holmes (ed.), *Virtual Politics: Identity and Community in Cyberspace*, London: Sage, pp. 1–25.

hooks, b. (1990) 'Choosing the Margin as a Space of Radical Openness', in *Yearning: Race, Gender, and Cultural Politics*, Boston, MA: South End Press, pp. 145–153.

Hopkins, L., Ewing, S., Meredyth, D., and Thomas, J. (2003) 'Machinery and Community: The Atherton Gardens Community Network', *Southern Review*, 36(1): 86–101.

Howard, A. (1999) 'Pacific-Based Virtual Communities: Rotuma on the World Wide Web', *The Contemporary Pacific*, 11(1): 160–175.

Howard, S. (1994) 'How Your Home Will Operate', *Business Review Weekly*, 25 April: 100.

Hughes, A. A., and Hughes, T. P. (2000a) 'Introduction', in A. A. Hughes and T. P. Hughes (eds), *Systems, Experts, and Computers: The Systems Approach in Management and Engineering, World War II and After*, Cambridge, MA: MIT Press, pp. 1–26.

———. (eds) (2000b) *Systems, Experts, and Computers: The Systems Approach in Management and Engineering, World War II and After*, Cambridge, MA: MIT Press.

Hughes, J. M., Michell, P. A., and Ramson, W. S. (eds) (1992) *The Australian Concise Oxford Dictionary*, Melbourne: Oxford University Press.

Hummon, D. M. (1990) *Commonplaces: Community Ideology and Identity in American Culture*, Albany, NY: State University of New York Press.

———. (1992) 'Community Attachment: Local Sentiment and Sense of Place', in I. Altman and S. M. Low (eds), *Place Attachment*, New York: Plenum Press, pp. 253–278.

Humphreys, L. (2005) 'Cellphones in Public: Social Interactions in a Wireless Era', *New Media and Society*, 7(6): 810–833.

Hunt, G. (1998) 'Architecture in the "Cybernetic Age"', *Architectural Design* (Architectural Design Profile No. 136: 'Architects in Cyberspace II'), 68(11/12), November–December: 53–55.

Hunter, D. (2003) 'Cyberspace as Place, and the Tragedy of the Digital Anticommons', *California Law Review*, 91(2), March: 439–519.

Huysman, M., and Wulf, V. (2005) 'The Role of Information Technology in Building and Sustaining the Relational Base of Communities', *The Information Society*, 21(2): 81–89.

Ibbett, R. N., and Topham, N. P. (1989) *Architecture of High Performance Computers—Volume 1*, Houndmills, Basingstoke, Hampshire: Macmillan Education.

Ihde, D. (1990) *Technology and the Lifeworld: From Garden to Earth*, Bloomington, IN: Indiana University Press.

Ito, M. (2003) 'Mobiles and the Appropriation of Place', *Receiver: Mobile Environment*, 8, <http://www.receiver.vodafone.com> (accessed 15 February 2006).

———. (2005) 'Mobile Phones, Japanese Youth, and the Re-Placement of Social Contact', in R. Ling and P. E. Pedersen (eds), *Mobile Communications: Renegotiation of the Social Sphere*, London: Springer, pp. 131–148.

Ito, M., and Okabe, D. (2005) 'Technosocial Situations: Emergent Structurings of Mobile Email Use', in M. Ito, D. Okabe, and M. Matsuda (eds), *Personal, Portable, and Pedestrian: Mobile Phones in Japanese Life*, Cambridge, MA: MIT Press, pp. 257–273.

Jabès, E. (1993) *The Book of Margins*, trans. R. Waldrop, Chicago, IL: University of Chicago Press.

Jackson, S. (2002) 'Review: *Systems, Experts, and Computers: The Systems Approach in Management and Engineering, World War II and After*, edited by A. C. Hughes and T. P. Hughes', *The Information Society*, 18(1): 69–70.

Jacobs, J. (1972) *The Death and Life of Great American Cities*, Harmondsworth, Middlesex: Penguin / Jonathan Cape.

James, I. (2006) *The Fragmentary Demand: An Introduction to the Philosophy of Jean-Luc Nancy*, Stanford, CA: Stanford University Press.

James, P., and Carkreek, F. (1997) 'This Abstract Body: From Embodied Symbolism to Techno-Disembodiment', in D. Holmes (ed.), *Virtual Politics: Identity and Community in Cyberspace*, London: Sage, pp. 107–124.

Jameson, F. (1991) *Postmodernism; or, the Cultural Logic of Late Capitalism*, London: Verso.

Jameson, F., and Miyoshi, M. (eds) (1998) *Cultures of Globalization*, Durham, NC: Duke University Press.

Jankowski, N. W. (2002) 'Creating Community with Media: History, Theories and Scientific Investigations', in L. A. Lievrouw and S. Livingstone (eds), *Handbook of New Media: Social Shaping and Consequences of ICTs*, London: Sage, pp. 34–49.

Jankowski, N. W., Van Selm, M., and Hollander, E. (2001) 'On Crafting a Study of Digital Community Networks', in L. Keeble and B. Loader (eds), *Community Informatics: Shaping Computer-Mediated Social Relations*, London: Routledge, pp. 101–117.

Jansz, J., and Martens, L. (2005) 'Gaming at a LAN Event: The Social Context of Playing Video Games', *New Media and Society*, 7(3): 333–355.

Jencks, C. (1995) *The Architecture of the Jumping Universe: A Polemic: How Complexity Science Is Changing Architecture and Culture*, London: Academy Editions.

Johnson, B. (1981) 'Translator's Introduction', in J. Derrida, *Dissemination*, trans. B. Johnson, Chicago, IL: The University of Chicago Press, pp. vii–xxxiii.

Johnson, L., and Lloyd, J. (2004) *Sentenced to Everyday Life: Feminism and the Housewife*, Oxford: Berg.

Johnson, S. (1997) *Interface Culture: How New Technology Transforms the Way We Create and Communicate*, San Francisco: HarperEdge.

Jonassen, C. T. (1959) 'Community Typology', in M. B. Sussman (ed.), *Community Structure and Analysis*, New York: Thomas Y. Crowell Company, pp. 15–36.

Jones, S. G. (1995) 'Introduction: From Where to Who Knows?', in S. G. Jones (ed.), *Cybersociety: Computer-Mediated Communication and Community*, Thousand Oaks: Sage, pp. 1–9.

———. (1997) 'The Internet and Its Social Landscape', in S. G. Jones (ed.), *Virtual Culture: Identity and Communication in Cybersociety*, London: Sage, pp. 7–35.

———. (1998) 'Information, Internet, and Community: Notes Toward an Understanding of Community in the Information Age', in S. G. Jones (ed.), *Cybersociety 2.0: Revisiting Computer-Mediated Communication and Community*, Thousand Oaks: Sage, pp. 1–34.

———. (ed.) (1999a) *Doing Internet Research: Critical Issues and Methods for Examining the Net*, London: Sage.

———. (1999b) 'Studying the Net: Intricacies and Issues', in S. G. Jones (ed.), *Doing Internet Research: Critical Issues and Methods for Examining the Net*, Thousand Oaks: Sage, pp. 1–27.

———. (2001) 'Understanding Micropolis and Compunity', in C. Ess and F. Sudweeks (eds), *Culture, Technology, Communication: Towards an Intercultural Global Village*, Albany, NY: State University of New York Press, pp. 53–66.

Joris, P. (1988) 'Translator's Preface', in M. Blanchot, *The Unavowable Community*, trans. P. Joris, Barrytown, NY: Station Hill Press, pp. xi–xxix.

Kaji-O'Grady, S. (2004) 'The Look of Information: Conceptual Art, Computers and Architecture in the 1970s', paper presented at 'Limits', XXIst annual conference of the Society of Architectural Historians Australia and New Zealand, Melbourne, Australia, 26–29 September.

Kamuf, P. (1991) 'On the Limit', in Miami Theory Collective (ed.), *Community at Loose Ends*, Minneapolis, MN: University of Minnesota Press, pp. 13–18.

Kaplan, S., Fitzpatrick, G., and Docherty, M. (2000) 'Stepping into Cooperative Buildings', paper presented at the First Australasian User Interface Conference (AUIC 2000): 'Stepping Out of Windows', Australian National University, Canberra, 31 January–3 February.

Karatani, K. (1995) *Architecture as Metaphor: Language, Number, Money*, trans. Sabu Kohso, Michael Speaks (ed.), Cambridge, MA: MIT Press.

Karim, K. H. (ed.) (2003) *The Media of Diaspora*, London: Routledge.

Kavanaugh, A. L., Reese, D. D., Carroll, J. M., and Rosson, M. B. (2005) 'Weak Ties in Networked Communities', *The Information Society*, 21(2): 119–131.

Keeble, L., and Loader, B. (eds) (2001) *Community Informatics: Shaping Computer-Mediated Social Relations*, London: Routledge.

Keenan, A. (1999) 'Home Sweet E-Home', *Sunday Age*: Property, 17 October: 1, 3.

Keith, M., and Pile, S. (eds) (1996) *Place and the Politics of Identity*, London: Routledge.

Keleman, M., and Smith, W. (2001) 'Community and its "Virtual" Promises: A Critique of Cyberlibertarian Rhetoric', *Information, Communication and Society*, 4(3): 370–387.

Keller, E. F. (2002) 'Marrying the Premodern to the Postmodern: Computers and Organisms after WWII', in D. Tofts, A. Jonson, and A. Cavallaro (eds), *Prefiguring Cyberculture: An Intellectual History*, Sydney / Cambridge, MA: Power Publications / MIT Press, pp. 52–65.

Kendall, L. (2002) *Hanging Out in the Virtual Pub: Masculinities and Relationships Online*, Berkeley, CA: University of California Press.

Kendrick, M. (1996) 'Cyberspace and the Technological Real', in R. Markley (ed.), *Virtual Realities and Their Discontents*, Baltimore, MD: The Johns Hopkins University Press, pp. 143–160.

Kipnis, J., and Leeser, T. (eds) (1987) *Chora L Works: Jacques Derrida and Peter Eisenman*, New York: The Monacelli Press.

Kitchin, R. (1998) *Cyberspace: The World in the Wires*, Chichester, West Sussex: John Wiley & Sons.

Kjaer, A., Halskov Madsen, K., and Graves Petersen, M. (2000) 'Methodological Challenges in the Study of Technology Use at Home', in A. Sloane and F. van Rijn (eds), *Home Informatics: Information, Technology and Society*, Boston, MA: Kluwer, pp. 45–60.

Knack, R. E. (2000) 'Hanging Out: Teens Search for the Perfect Public Place', *Planning*, 66(8), August: 4–9.

Koku, E., Nazer, N., and Wellman, B. (2001) 'Netting Scholars: Online and Offline', *American Behavioral Scientist*, 44(10): 1752–1774.

Kolko, B., and Reid, E. (1998) 'Dissolution and Fragmentation: Problems in on-Line Communities', in S. G. Jones (ed.), *Cybersociety 2.0: Revisiting Computer-Mediated Communication and Community*, Thousand Oaks: Sage, pp. 212–229.

Kollock, P. (1999) 'The Economies of Online Cooperation: Gifts and Public Goods in Cyberspace', in M. A. Smith and P. Kollock (eds), *Communities in Cyberspace*, London: Routledge, pp. 220–239.

Kollock, P., and Smith, M. A. (1999) 'Communities in Cyberspace', in M. A. Smith and P. Kollock (eds), *Communities in Cyberspace*, London: Routledge, pp. 3–25.

König, R. (1968) *The Community*, trans. E. Fitzgerald, London: Routledge & Kegan Paul Ltd.

Korhonen, K. (2006) 'Textual Communities: Nancy, Blanchot, Derrida', *Culture Machine*, 8, <http://www.culturemachine.net/index.php/cm/article/view/35/43> (accessed 27 November 2010).

Kraidy, M. M. (2001) 'From Imperialism to Glocalization: A Theoretical Framework for the Information Age', in B. Ebo (ed.), *Cyberimperialism?: Global Relations in the New Electronic Frontier*, Westport, CT: Praeger, pp. 27–42.

Kumar, K. (1987) *Utopia and Anti-Utopia in Modern Times*, Oxford: Basil Blackwell.

Lab—Bates + Davidson (1996) 'Architecture After Geometry', *Architectural Design* (Architectural Design Profile No. 123: 'Integrating Architecture'), 66(9/10): 66–71.

Lahtinen, H. (2000) 'On Purchasing a Home Computer', in A. Sloane and F. van Rijn (eds), *Home Informatics: Information, Technology and Society*, Boston, MA: Kluwer, pp. 191–215.

Lakoff, G., and Johnson, M. (1980) *Metaphors We Live By*, Chicago, IL: The University of Chicago Press.

Lally, E. (2002) *At Home with Computers*, Oxford: Berg.

Landau, R. (1984) 'A Philosophy of Enabling', in C. Price, *Cedric Price, Works II*, London: Architectural Association, pp. 9–15.

Lash, S. (1999) *Another Modernity, a Different Rationality*, Oxford: Blackwell.

Laurier, E. (2001) 'Why People Say Where They Are During Mobile Phone Calls', *Environment and Planning D: Society and Space*, 19: 485–504.

Laurier, E., Whyte, A., and Buckner, K. (2002) 'Neighbouring as an Occasioned Activity: "Finding a Lost Cat"', *Space and Culture*, 5(4), November: 346–367.

Lawrence, R. J. (1987) 'What Makes a House a Home?', *Environment and Behavior*, 19(2), March: 154–168.

Le Corbusier (1956) *The Modulor: A Harmonious Measure to the Human Scale Universally Applicable to Architecture and Mechanics*, trans. P. De Francia and A. Bostock, London: Faber and Faber.

——. (1958) *Modulor 2 1955 (Let the User Speak Next): Continuation of 'the Modulor' 1948*, trans. P. De Francia and A. Bostock, London: Faber and Faber.

Leccese, M., and McCormick K. (eds) (2000) *Charter of the New Urbanism: Region / Neighbourhood, District, and Corridor / Block, Street, and Building*, New York: McGraw-Hill.

Lee, P. M. (1998) 'On the Holes of History: Gordon Matta-Clark's Work in Paris', *October*, 85, Summer: 65–89.

Lefebvre, H. (2000) *The Production of Space*, trans. D. Nicholson-Smith, Oxford: Blackwell.

——. (2004) *Rhythmanalysis: Space, Time and Everyday Life*, trans. Stuart Elden and Gerald Moore, London: Continuum.

Lemos, A. (1996) 'The Labyrinth of Minitel', in R. Shields (ed.), *Cultures of Internet: Virtual Spaces, Real Histories, Living Bodies*, London: Sage, pp. 33–48.

Leonard, G. J. (1994) *Into the Light of Things: Art of the Commonplace from Wordsworth to John Cage*, Chicago, IL: University of Chicago Press.

Levinson, P. (2003) *Realspace: The Fate of Physical Presence in the Digital Age, On and Off the Planet*, London: Routledge.

Lévi-Strauss, C. (1966) *The Savage Mind*, trans. G. Weidenfeld and Nicolson Ltd, London: Weidenfeld and Nicolson.

Levitas, R. (1993) 'The Future of Thinking About the Future', in J. Bird, B. Curtis, T. Putnam, G. Robertson and L. Tickner (eds), *Mapping the Futures: Local Cultures, Global Change*, London: Routledge, pp. 257–266.

Lévy, P. (1998) *Becoming Virtual: Reality in the Digital Age*, trans. R. Bononno, New York: Plenum Trade.

Lewallen, C. M., and Seid, S. (eds) (2004) *Ant Farm 1968–1978*, Berkeley, CA: University of California Press / Berkeley Art Museum and Pacific Film Archive.

Lewis, T. (2001) 'Locating Community in the Social: Reorienting Internet Research', in H. Brown, G. Lovink, H. Merrick, N. Rossiter, D. Teh and M. Willson (eds), *Politics of a Digital Present: An Inventory of Australian Net Culture*, Melbourne: Fibreculture Publications, pp. 233–242.

Licoppe, C. (2004) '"Connected" Presence: The Emergence of a New Repertoire for Managing Social Relationships in a Changing Communication Technoscape', *Environment and Planning D: Society and Space*, 22(1): 135–156.

Lim, S. S., and Chung, L.-Y. (2004) '*The Dance of Life* (Digital Remix): The Impact of Mobile Communication on Time Use', *Media Asia*, 31(1): 37–43.

Ling, R. (2000) 'Direct and Mediated Interaction in the Maintenance of Social Relationships', in A. Sloane and F. van Rijn (eds), *Home Informatics and Telematics: Information, Technology and Society*, Boston, MA: Kluwer Academic Publishers, pp. 61–86.

Ling, R., and Haddon, L. (2003) 'Mobile Telephony, Mobility and the Coordination of Everyday Life', in J. Katz (ed.), *Machines that Become Us: The Social Context of Personal Communication Technology*, New Brunswick, NJ: Transaction Publishers, pp. 245–265.

Lingis, A. (1994) *The Community of Those Who Have Nothing in Common*, Bloomington, IN: Indiana University Press.

Littlewood, J. (1964) 'A Laboratory of Fun', *New Scientist*, 22(391), 14 May: 432–433.

Liu, A. (1989) 'The Power of Formalism: The New Historicism', *ELH*, 56(4), Winter: 721–771.

Lobsinger, M. L. (2000a) 'Cedric Price: An Architecture of Performance', *Daidalos*, 74, October: 22–29.

———. (2000b) 'Cybernetic Theory and the Architecture of Performance: Cedric Price's Fun Palace', in S. Williams Goldhagen and R. Legault (eds), *Anxious Modernisms: Experimentation in Postwar Architectural Culture*, Montreal, Canada / Cambridge, MA: Canadian Centre for Architecture / MIT Press, pp. 119–139.

Lombard, M., and Ditton, T. (1997) 'At the Heart of It All: The Concept of Presence', *Journal of Computer-Mediated Communication*, 3(2), September, <http://jcmc.indiana.edu/vol3/issue2/lombard.html> (accessed 27 November 2010).

Loomis, C. P. (1963) 'Introduction: Tönnies and His Relation to Sociology—Orientation of *Gemeinschaft und Gesellschaft*', in F. Tönnies, *Community and Society (Gemeinschaft und Gesellschaft)*, trans. C. P. Loomis, New York: Harper Row, pp. 1–11.

Lord, C., Michels, D., and Schreier, C. (2004) 'Ant Farm Timeline', in C. M. Lewallen and S. Seid (eds), *Ant Farm 1968–1978*, Berkeley, CA: University of California Press / Berkeley Art Museum and Pacific Film Archive, pp. 88–149.

Lovink, G. (2002) *Dark Fiber: Tracking Critical Internet Culture*, Cambridge, MA: MIT Press.

Lucy, N. (1995) *Debating Derrida*, Melbourne: Melbourne University Press.

———. (1997) *Postmodern Literary Theory: An Introduction*, Oxford: Blackwell.

———. (2000) *Postmodern Literary Theory: An Anthology*, Oxford: Blackwell.

———. (2001) *Beyond Semiotics: Text, Culture and Technology*, London: Continuum.

———. (2004) *A Derrida Dictionary*, Oxford: Blackwell.

Lukermann, F. E. (1964) 'Geography as a Formal Intellectual Discipline and the Way in Which It Contributes to Human Knowledge', *Canadian Geographer*, 8(4): 167–172.

Lüschen, G., and Stone, G. P. (eds) (1977) *Herman Schmalenbach: On Society and Experience*, Chicago, IL: University of Chicago Press.

Lyon, D. (1997) 'Cyberspace Sociality: Controversies over Computer-Mediated Relationships', in B. D. Loader (ed.), *The Governance of Cyberspace: Politics, Technology and Global Restructuring*, London: Routledge, pp. 23–37.

Macherey, P. (2006) *A Theory of Literary Production*, trans. G. Wall, London: Routledge.

Madsen, V. (1996) 'Virtual Light', *21C*, 1: 46–49.

Malpas, J. (1999) *Place and Experience: A Philosophical Topography*, Cambridge, UK: Cambridge University Press.

————. (2006) *Heidegger's Topology: Being, Place, World*, Cambridge, MA: MIT Press.

Maniaque, C. (2004) 'Searching for Energy', in C. M. Lewallen and S. Seid (eds), *Ant Farm 1968–1978*, Berkeley, CA: University of California Press / Berkeley Art Museum and Pacific Film Archive, pp. 14–21.

Manning, P. (1973) 'Architecture, Computers and the Future, Report # 20', Halifax, Nova Scotia: Nova Scotia Technical College.

Manovich, L. (2001) *The Language of New Media*, Cambridge, MA: MIT Press.

Mantovani, G., and Riva, G. (1998) '"Real" Presence: How Different Ontologies Generate Different Criteria for Presence, Telepresence and Virtual Presence', *Presence: Teleoperators and Virtual Environments*, 1(1): 540–550.

Marchbank, T. (2004) 'Intense Flows: Flashmobbing, Rush Capital and the Swarming of Space', *Philament*, 4, August, <http://www.arts.usyd.edu.au/publications/philament/issue4_Critique_Marchbank.htm> (accessed 27 November 2010).

Marshall, J. (2000) 'Living Online: Categories, Communication and Control', unpublished PhD thesis, Department of Anthropology, University of Sydney.

————. (2001) 'Cyber-Space, or Cyber-Topos: The Creation of Online Space', *Social Analysis*, 45(1), April: 81–102.

Martin, E. (ed.) (1994) *Architecture as a Translation of Music*, New York: Princeton Architectural Press.

Martin, J. (1981) *Telematic Society: A Challenge for Tomorrow*, Englewood Cliffs, NJ: Prentice-Hall.

Marvin, C. (1988) *When Old Technologies Were New: Thinking About Electric Communication in the Late Nineteenth Century*, New York: Oxford University Press.

Marvin, S. (1997) 'Telecommunications and the Urban Environment: Electronic and Physical Links', in P. Droege (ed.), *Intelligent Environments: Spatial Aspects of the Information Revolution*, Amsterdam: Elsevier Science, pp. 179–197.

Marx, G. T. (1997) 'The Declining Significance of Traditional Borders (and the Appearance of New Borders) in an Age of High Technology', in P. Droege (ed.), *Intelligent Environments: Spatial Aspects of the Information Revolution*, Amsterdam: Elsevier Science, pp. 484–493.

Massey, D. (1991) 'A Global Sense of Place', *Marxism Today*, June: 24–26 & 28–29.

————. (1992a) 'A Place Called Home?', *New Formations*, 17, Summer: 3–15.

————. (1992b) 'Politics and Space/Time', *New Left Review*, 196, November/December: 65–84.

————. (1999) 'Negotiating Disciplinary Boundaries', *Current Sociology*, 47(4), October: 5–12.

————. (2002) 'Don't Let's Counterpose Place and Space', *Development*, 45(1), March: 24–25.

————. (2005) *For Space*, London: Sage.

Mathews, S. (2006) 'The Fun Palace as Virtual Architecture: Cedric Price and the Practices of Indeterminacy', *Journal of Architectural Education*, 59(3), February: 39–48.

May, T. (1997) *Reconsidering Difference: Nancy, Derrida, Levinas, and Deleuze*, University Park, PA: Pennsylvania State University Press.

McCormack, P. (1998) 'Communicating Community: Past and Present', *M/C: A Journal of Media and Culture*, 1(1), <http://www.uq.edu.au/mc/9807/comp.html> (accessed 2 June 2000).

McCullough, M. (2004) *Digital Ground: Architecture, Pervasive Computing and Environmental Knowing*, Cambridge, MA: MIT Press.

McDonough, M. (2003) 'The Smart House Examined', in J. G. Trulove (ed.), *The Smart House*, New York: HDI, pp. 6–13.

McGuire, M. (1995) 'A-Symmetry City', *Architectural Design* (Architectural Design Profile No. 118: 'Architects in Cyberspace'), 65(11/12), November–December: 52–57.

McKinney, J. C., and Loomis, C. P. (1963) 'Introduction: The Application of Gemeinschaft and Gesellschaft as Related to Other Typologies', in F. Tönnies, *Community and Society (Gemeinschaft und Gesellschaft)*, trans and ed. C. P. Loomis, New York: Harper and Row, pp. 12–29.

McLuhan, M., and Fiore, Q. (1989) *The Medium Is the Massage*, New York: Touchstone.

McQuire, S. (1997a) 'The Uncanny Home, Or Living On-Line with Others', in P. Droege (ed.), *Intelligent Environments: Spatial Aspects of the Information Revolution*, Amsterdam: Elsevier Science, pp. 682–709.

———. (1997b) 'The Uncanny Home: Television, Transparency and Overexposure', *Paradoxa*, 3(3–4): 527–538.

———. (2002) 'Space For Rent in the Last Suburb', in D. Tofts, A. Jonson, and A. Cavallaro (eds), *Prefiguring Cyberculture: An Intellectual History*: Sydney / Cambridge, MA: Power Publications / MIT Press, pp. 166–178.

———. (2003) 'From Glass Architecture to *Big Brother*: Scenes from a Cultural History of Transparency', *Cultural Studies Review*, 9(1), May: 103–123.

Memarzia, K. (1997) 'Towards the Definition and Applications of Digital Architecture', unpublished thesis, Department of Architectural Studies, University of Sheffield, <http://bip.concept.se/user/kmemarzia/dis97.htm#_Toc389430053> (accessed 16 July 1998).

Meyer, E., Gyalong, T., Overney, R. M., and Dransfield, K. (eds) (1999) *Nanoscience: Friction and Rheology on the Nanometer Scale*, Singapore: World Scientific.

Meyrowitz, J. (1985) *No Sense of Place: The Impact of Electronic Media on Social Behavior*, New York: Oxford University Press.

Miami Theory Collective (ed.) (1991) *Community at Loose Ends*, Minneapolis, MN: University of Minnesota Press.

Miles, A. (2002) 'Busy Day', *Videoblog::Vog 2.0* [weblog], 18 September (2002), <http://hypertext.rmit.edu.au/vog/vlog/archive/2002/92002.html> (accessed 27 November 2010).

Miller, D., and Slater, D. (2000) *The Internet: An Ethnographic Approach*, London: Routledge.

Milne, E. (2000) 'Vicious Circles: Metaphor and the Historiography of Cyberspace', *Social Semiotics*, 10(1): 99–108.

———. (2007) '"Dragging Her Dirt All Over the Net": Presence, Intimacy, Materiality V1.0', *Transforming Cultures eJournal*, 2(2), December, <http://epress.lib.uts.edu.au/ojs/index.php/TfC/article/view/636/564> (accessed 23 November 2010).

———. (2010) *Letters, Postcards, Email: Technologies of Presence*, New York: Routledge.

Minsky, M. (1980) 'Telepresence', *Omni*, June: 45–51.

Misa, T. J. (2003) 'The Compelling Tangle of Modernity and Technology', in T. J. Misa, P. Brey, and A. Feenberg (eds), *Modernity and Technology*, Cambridge, MA: MIT Press, pp. 1–30.

Mitchell, W. J. (1995) *City of Bits: Space, Place, and the Infobahn*, Cambridge, MA: MIT Press.

———. (1999a) *E-Topia: "Urban Life, Jim—But Not as We Know It"*, Cambridge, MA: MIT Press.

———. (1999b) 'Foreword: Who Put the Space in Cyberspace?', in P. Anders (ed.), *Envisioning Cyberspace: Designing 3D Electronic Spaces*, New York: McGraw-Hill, pp. x–xii.

——. (2000) 'Replacing Place', in P. Lunenfeld (ed.), *The Digital Dialectic: New Essays on New Media*, Cambridge, MA: MIT Press, pp. 112–128.

——. (2005) *Placing Words: Symbols, Space, and the City*, Cambridge, MA: MIT Press.

Moran, J. (2002) *Interdisciplinarity*, London: Routledge.

Morin, M.-E. (2006) 'Putting Community Under Erasure: Derrida and Nancy on the Plurality of Singularities', *Culture Machine*, 8, <http://www.culturemachine. net/index.php/cm/article/viewArticle/37/45> (accessed 27 November 2010).

Morley, D. (2000) *Home Territories: Media, Mobility and Identity*, London: Routledge.

——. (2003) 'What's "Home" Got to Do with It?: Contradictory Dynamics in the Domestication of Technology and the Dislocation of Domesticity', *European Journal of Cultural Studies*, 6(4): 435–458.

——. (2007) *Media, Modernity and Technology: The Geography of the New*, London: Routledge.

Morley, D., and Robins, K. (eds) (1995), *Spaces of Identity: Global Media, Electronic Landscapes and Cultural Boundaries*, London: Routledge.

Morningstar, C., and Farmer, F. R. (1992) 'The Lessons of Lucasfilm's *Habitat*', in M. Benedikt (ed.), *Cyberspace: First Steps*, Cambridge, MA: MIT Press, pp. 273–301.

Morris, M. (1996) 'Crazy Talk Is Not Enough', *Environment and Planning D: Society and Space*, 14(4): 384–394.

Morse, M. (1990) 'An Ontology of Everyday Distraction: The Freeway, the Mall, and Television', in P. Mellencamp (ed.), *Logics of Television: Essays in Cultural Criticism*, Bloomington, IN / London: Indiana University Press / BFI Publishing, pp. 193–221.

Mumford, L. (1979) *The City in History: Its Origins, Its Transformations, and Its Prospects*, Harmondsworth, Middlesex: Penguin / Martin Secker & Warburg.

Muñoz González, B. (2005) 'Topophilia and Topophobia: The Home as an Evocative Place of Contradictory Emotions', *Space and Culture*, 8(2): 193–215.

Murdock, G., Harmann, P., and Gray, P. (1994) 'Contextualizing Home Computing: Resources and Practices', in R. Silverstone and E. Hirsch (eds), *Consuming Technologies: Media and Information in Domestic Spaces*, London: Routledge, pp. 146–160.

Murphie, A. (2004) '::Fibreculture:: Virtuosity, Processual Democracy and Organised Networks', Fibreculture list, 22 September, <http://lists.myspinach. org/archives/fibreculture/2004-September/003974.html> (accessed 23 September 2004).

Murphie, A., and Potts, J. (2003) *Culture and Technology*, Houndsmills, Basingstoke, Hampshire, UK: Palgrave Macmillan.

Nancy, J.-L. (1991a) 'Introduction', in E. Cadava, P. Connor, and J.-L. Nancy (eds), *Who Comes After the Subject?* New York: Routledge, pp. 1–8.

——. (1991b) *The Inoperative Community*, trans. P. Connor, L. Garbus, M. Holland, and S. Sawhney, Minneapolis, MN: University of Minnesota Press.

——. (1991c) 'Of Being-in-Common', trans. J. Creech, in Miami Theory Collective (ed.), *Community at Loose Ends*, Minneapolis, MN: University of Minnesota Press, pp. 1–12.

——. (1992) 'La Comparution/The *Compearance* from the Existence of "Communism" to the Community of "Existence"', trans. T. B. Strong, *Political Theory*, 20(3), August: 371–399.

——. (2000) *Being Singular Plural*, trans. R. D. Richardson and A. E. O'Byrne, Stanford, CA: Stanford University Press.

——. (2003) 'The Confronted Community', trans. A. Macdonald, *Postcolonial Studies*, 6(1): 23–36.

Nansen, B., Arnold, M., Gibbs, M., and Davis, H. (2009) 'Domestic Orchestration: Rhythms in the Mediated Home', *Time & Society*, 18(2): 181–207.

Negroponte, N. (1970) *The Architecture Machine*, Cambridge, MA: MIT Press.

———. (1975a) 'Foreword to the English Language Edition', in Y. Friedman, *Toward a Scientific Architecture*, trans. C. Lang, Cambridge, MA: MIT Press, pp. ix–x.

———. (ed.) (1975b) *Reflections on Computer Aids to Design and Architecture*, New York: Petrocelli/Charter.

———. (1975c) *Soft Architecture Machines*, Cambridge, MA: MIT Press.

———. (1995) *Being Digital*, Cambridge, MA: MIT Press.

Negus, M. (1998) 'Corporeal Architecture: Some Current Positions', *Architectural Theory Review*, 3(2): 112–122.

Nelson, L. (1955) *Rural Sociology*, 2nd edition, New York: American Book Company.

Newman, W. M. (1966) 'An Experimental Program for Architectural Design', *Computer Journal*, May: 21–26.

Nguyen, N. H. C. (2005) *Voyage of Hope: Vietnamese Australian Women's Narratives*, Altona, Victoria: Common Ground.

Nicholson, J. A. (2005) 'Flash! Mobs in the Age of Mobile Connectivity', *Fibreculture Journal*, 6, <http://six.fibreculturejournal.org/> (accessed 27 November 2010).

Nippert-Eng, C. E. (1996) *Home and Work: Negotiating Boundaries through Everyday Life*, Chicago, IL: The University of Chicago Press.

Nisbet, R. A. (1953) *The Quest for Community: A Study in the Ethics of Order and Freedom*, New York: Oxford University Press.

———. (1967) *The Sociological Tradition*, London: Heinemann.

———. (1969) *Social Change and History; Aspects of the Western Theory of Development*, New York: Oxford University Press.

Nixon, M. (1996) 'The Architecture of Change: De Recombinant Architectura', *21C*, 1: 38–45.

Norberg-Schulz, C. (1976) 'The Phenomenon of Place', *Architectural Association Quarterly*, 8(4): 3–10.

———. (1980) *Genius Loci: Towards a Phenomenology of Architecture*, London: Academy Editions.

Norwich, W. (1999) 'Open House', *House and Garden*, July: 56, 59–60.

Novak, M. (1992) 'Liquid Architectures in Cyberspace', in M. Benedikt (ed.), *Cyberspace: First Steps*, Cambridge, MA: The MIT Press, pp. 225–254.

———. (1994) 'Computation and Composition', in E. Martin (ed.), *Architecture as a Translation of Music*, New York: Princeton Architectural Press, pp. 64–67.

———. (1995) 'Transmitting Architecture: transTerraFirma/TidsvagNoll v2.0', *Architectural Design* (Architectural Design Profile No. 118: 'Architects in Cyberspace'), 65(11/12), November-December: 42–47.

———. (1996) 'transArchitecture', *Telepolis: Magazin der Netzkultur* ('Architektur' special edition), <http://www.heise.de/tp/r4/artikel/6/6069/2.html> (accessed 27 November 2010).

———. (1997) 'Cognitive Cities: Intelligence, Environment and Space', in P. Droege (ed.), *Intelligent Environments: Spatial Aspects of the Information Revolution*, Amsterdam: Elsevier Science, pp. 386–419.

———. (1998) 'Next Babylon, Soft Babylon: (trans)Architecture is an Algorithm to Play In', *Architectural Design* (Architectural Design Profile No. 136: 'Architects in Cyberspace II'), 68(11/12), November-December: 21–29.

Nunes, M. (1997) 'What Space Is Cyberspace?: The Internet and Virtuality', in D. Holmes (ed.), *Virtual Politics: Identity and Community in Cyberspace*, Thousand Oaks: Sage, pp. 163–178.

O'Brien, J. (1999) 'Writing in the Body: Gender (Re)Production in Online Interaction', in M. A. Smith and P. Kollock (eds), *Communities in Cyberspace*, London: Routledge, pp. 76–104.

Obrist, H.-U., and Virilio, P. (n.d.) 'Interview with Paul Virilio', <http://www.ljudmila.org/scca/urbanaria/txt/e/pvirilio.htm> (accessed 27 November 2010).

O'Connor, A. (1989) *Raymond Williams: Writing, Culture, Politics*, Oxford: Basil Blackwell.

OECD (1998) *21ˢᵗ Century Technologies: Promises and Perils of a Dynamic Future*, Paris: Organisation for Economic Co-Operation and Development.

O'Gorman, M. (2004) 'American ~~Beauty~~ Busted: Necromedia and Domestic Discipline', *SubStance*, 33(3): 34–51.

Olsen, L. (1991) 'The Shadow of Spirit in William Gibson's Matrix Trilogy', *Extrapolation*, 32(3): 278–289.

Olson, K. (2005) 'Cyberspace as Place and the Limits of Metaphor', *Convergence*, 11(1): 10–18.

Ostwald, M. (1996) 'Understanding Cyberspace: Learning from Luna Park', *Architecture Australia*, March/April: 84–87.

Ostwald, M. J., and Moore, R. J. (1998) *Disjecta Membra: Architecture and the Loss of the Body*, Sydney: Archadia Press.

O'Toole, M. (1994) *The Language of Displayed Art*, London: Leicester University Press.

P. R. B. (1957) 'The Adaptable House', *The Architectural Review*, 121(721), February: 84 & 130.

Parrish, R. (2002) 'The Changing Nature of Community', *Strategies: Journal of Theory, Culture, and Politics*, 15(2), November: 259–284.

Pask, G. (1960) *An Approach to Cybernetics*, New York: Harper.

———. (1969) 'The Architectural Relevance of Cybernetics', *Architectural Design* ('Thinking About Architecture & Planning'), 39(7/6), September: 494–496.

Pawley, M. (1990) *Theory and Design in the Second Machine Age*, Oxford: Basil Blackwell.

———. (1998) *Terminal Architecture*, London: Reaktion Books.

Pearce, M. (1995) 'From Urb to Bit', *Architectural Design* (Architectural Design Profile No. 118: 'Architects in Cyberspace'), 65(11/12), November–December: 7.

Pelikan Strau, N. (1997) 'The Fourth Blow to Narcissism and the Internet', *Literature and Psychology*, 43(1–2), Spring–Summer: 96–110.

Perec, G. (1999) *Species of Spaces and Other Pieces*, trans. and ed. J. Sturrock, Harmondsworth, Middlesex: Penguin.

Philipson, G. (2006) 'Your Tired, Your Poor—Your Mobiles', *Age*: Issues, Tuesday 19 September: 5.

Pingree, G. B., and Gitelman, L. (2003), 'Introduction: What's New About New Media?', in L. Gitelman and G. B. Pingree (eds), *New Media, 1740–1915*, Cambridge, MA: MIT Press, pp. xi–xxii.

Plant, R. (1974) *Community and Ideology: An Essay in Applied Social Philosophy*, London: Routledge.

Plunkett, S. (1994) 'At Home in the Future', *Business Review Weekly*, 25 April: 98–100.

Polano, S. (1988) 'De Stijl/Architecture = Nieuwe Beelding', trans. J. Daley, in M. Friedman (ed.), *De Stijl: 1917–1931—Visions of Utopia*, 3ʳᵈ impression, Oxford: Phaidon, pp. 87–97.

Poster, M. (1995) 'Postmodern Virtualities', in M. Featherstone and R. Burrows (eds), *Cyberspace, Cyberbodies, Cyberpunk: Cultures of Technological Embodiment*, London: Sage, pp. 79–95.

———. (1998) 'Virtual Ethnicity: Tribal Identity in an Age of Global Communications', in S. G. Jones, *Cybersociety 2.0: Revisiting Computer-Mediated Communication and Community*, Thousand Oaks: Sage, pp. 184–211.

Price, C. (1964) 'A Laboratory of Fun', *New Scientist*, 22(391), 14 May: 433.

———. (1966a) 'Life-Conditioning', *Architectural Design*, XXXVI(10), October: 483.

———. (1966b) 'PTb: Potteries Thinkbelt: A Plan for an Advanced Educational Industry in North Staffordshire', *Architectural Design*, XXXVI(10), October: 483–497.

———. (1984) *Cedric Price, Works II*, London: Architectural Association.

———. (2003) *Cedric Price: The Square Book*, Chicester, West Sussex: Wiley-Academy.

Rabeneck, A. (1976) 'Whatever Happened to the Systems Approach?' *Architectural Design*, XLVI(5), May: 298–303.

Rahim, A. (2000) 'Systemic Delay: Breaking the Mold', *Architectural Design*, 70(3), June: 6–8.

Rainie, L., and Bell, P. (2004) 'The Numbers that Count', *New Media and Society*, 6(1): 44–54.

Rapaport, H. (2003) *Later Derrida: Reading the Recent Work*, New York: Routledge.

Rasmussen, S. E. (1999) *Experiencing Architecture*, 20th reprint, Cambridge, MA: MIT Press.

Rauch, A. (1996) 'Culture's Hieroglyph in Benjamin and Novalis: A Matter of Feeling', *The Germanic Review: Literature, Culture, Theory*, 71(4): 253–266.

Real, M. (1996) *Exploring Media Culture: A Guide*, Thousand Oaks: Sage.

Reid, E. (1995) 'Virtual Worlds: Culture and Imagination', in S. G. Jones (ed.), *Cybersociety: Computer-Mediated Communication and Community*, London: Sage, pp. 164–183.

———. (1999) 'Hierarchy and Power: Social Control in Cyberspace', in M. A. Smith and P. Kollock (eds), *Communities in Cyberspace*, London: Routledge, pp. 107–133.

Relph, E. (1986) *Place and Placelessness*, reprint and 3rd imprint, London: Pion.

Reynolds, R. A. (1980) *Computer Methods for Architects*, London: Butterworths.

Rheingold, H. (1993) 'A Slice of Life in My Virtual Community', in L. M. Harasim (ed.), *Global Networks: Computers and International Communication*, Cambridge, MA: MIT Press, pp. 57–80.

———. (1994) *Virtual Community: Finding Connection in a Computerized World*, London: Secker and Warburg.

———. (1999) 'Virtual Community: Another Metaphor', *Whole Earth*, Fall: 18.

———. (2002) *Smart Mobs: The Next Social Revolution*, Cambridge, MA: Perseus.

———. (2008) 'Mobile Media and Political Collective Action', in J. E. Katz (ed.), *Handbook of Mobile Communication Studies*, Cambridge, MA: MIT Press, pp. 225–239.

Richardson, I. (2005) 'Mobile Technosoma: Some Phenomenological Reflections on Itinerant Media Devices', *Fibreculture Journal*, 6, <http://six.fibreculture-journal.org/> (accessed 27 November 2010).

Richardon, I., and Wilken, R. (2009) 'Haptic Vision, Footwork, Place-making: A Peripatetic Phenomenology of the Mobile Phone Pedestrian', *Second Nature: International Journal of Creative Media*, 1(2), <http://secondnature.rmit.edu.au/index.php/2ndnature/article/view/121/35> (accessed 27 November 2010).

Ricoeur, P. (1977) *The Rule of Metaphor: Multi-Disciplinary Studies of the Creation of Meaning in Language*, trans. R. Czerny, K. McLaughlin, and J. Costello, Toronto: University of Toronto Press.

Riewoldt, O. (1997) *Intelligent Spaces: Architecture for the Information Age*, London: Laurence King.

Robertson, R. (1995) 'Glocalization: Time-Space and Homogeneity-Heterogeneity', in M. Featherstone, S. Lash, and R. Robertson (eds), *Global Modernities*, London: Sage, pp. 25–44.

Robins, K. (1999) 'Against Virtual Community: For a Politics of Difference', *Angelaki: Journal of the Theoretical Humanities*, 4: 163–170.

———. (2000) 'Cyberspace and the World We Live In', in D. Bell and B. Kennedy (eds), *The Cybercultures Reader*, London: Routledge, pp. 77–95.

Robins, K., and Cornford, J. (1990) 'Bringing It All Back Home', *Futures*, 22(8), October: 870–879.

Robins, K., and Webster, F. (1999) *Times of the Technoculture: From the Information Society to the Virtual Life*, London: Routledge.

Rodgers, J. (2004) 'Doreen Massey: Space, Relations, Communications', *Information, Communication & Society*, 7(2), June: 273–291.

Romero, C. (1998) 'Vortex 2000', *Architectural Design* (Architectural Design Profile No. 136: 'Architects in Cyberspace II'), 68(11/12), November–December: 46–51.

Rooksby, E. (2002) *E-Mail and Ethics: Style and Ethical Relations in Computer-Mediated Communication*, London: Routledge.

Rössler, P. (2001) 'Between Online Heaven and Cyberhell: The Framing of "The Internet" by Traditional Media Coverage in Germany', *New Media and Society*, 3(1): 49–66.

Rouner, L. S. (1991) 'Introduction', in L. S. Rouner (ed.), *On Community*, Notre Dame, IN: University of Notre Dame Press, pp. 1–11.

Rushkoff, D. (1994) *Cyberia: Life in the Trenches of Cyberspace*, London: Harper Collins.

Russell, B. (1980) *The Problems of Philosophy*, Oxford: Oxford University Press.

Sadler, S. (1998) *The Situationist City*, Cambridge, MA: MIT Press.

Sampson, A. (1996) 'Inside Out: Clinical Procedure Relating to Caesarean Section as Practised on Bodies, Objects and Buildings', *Architectural Design* (Architectural Design Profile No. 123: 'Integrating Architecture'), 66(9/10): 82–85.

Sardar, Z. (1996) 'alt.civilizations.faq: Cyberspace as the Darker Side of the West', in Z. Sardar and J. R. Ravetz (eds), *Cyberfutures: Culture and Politics on the Information Superhighway*, New York: New York University Press, pp. 14–41.

Sassen, S. (2000) 'Spatialities and Temporalities of the Global: Elements for a Theorization', *Public Culture*, 12(1): 215–232.

Savat, D., and Poster, M. (eds) (2009) *Deleuze and New Technology*, Edinburgh: Edinburgh University Press.

Schiano, D. J., and White, S. (1998) 'The First Noble Truth of Cyberspace: People Are People (Even When They MOO)', in C.-M. Karat, A. Lund, J. Coutaz, and J. Karat (eds), *Human Factors in Computing Systems: CHI '98 Conference Proceedings*, New York: The Association for Computing Machinery, pp. 352–359.

Schmalenbach, H. (1922) 'Die Soziologische Kategorie Des Bundes', *Die Dioskuren*, 1: 35–105.

———. (1961) 'The Sociological Category of Communion', trans. K. D. Naegele and G. P. Stone, in T. Parsons, E. Shils, K. Naegele, and J. R. Pitts (eds), *Theories of Society: Foundations of Modern Sociological Theory, Volume 1*, New York: The Free Press of Glencoe, pp. 331–347.

———. (1977) 'Communion—A Sociological Category', trans. G. Luschen and G. P. Stone, in G. Luschen and G. P. Stone (eds), *Herman Schmalenbach: On Society and Experience*, Chicago: University of Chicago Press, pp. 64–125.

Schmitz, J. (1997) 'Structural Relations, Electronic Media, and Social Change: The Public Electronic Network and the Homeless', in S. G. Jones (ed.), *Virtual Culture: Identity and Communication in Cybersociety*, London: Sage, pp. 80–101.

Schofield, J. (2004) 'Every Breath You Take', *Age*: Green Guide, 2 September: 2 & 6.

Scholte, J. A. (2000) *Globalization: A Critical Introduction*, Houndsmills, Basingstoke, Hampshire: Macmillan.

Scriver, P. (2002) 'On Place', in S. Akkach (ed.), *De-Placing Difference: Architecture, Culture and Imaginative Geography*, Adelaide: Centre for Asian and Middle Eastern Architecture, The University of Adelaide, pp. 3–8.

Sconce, J. (2000) *Haunted Media: Electronic Presence from Telegraphy to Television*, Durham, NC: Duke University Press.

Seabrook, J. (1997) *Deeper: A Two-Year Odyssey in Cyberspace*, London: Faber and Faber.

Secomb, L. (2003a) 'Interrupting Mythic Community', *Cultural Studies Review*, 9(1), May: 85–100.

——. (2003b) 'Introduction', *Cultural Studies Review*, 9(1), May: 9–11.

Segaller, S. (1998) *Nerds$^{2.0.1}$: A Brief History of the Internet*, New York: TV Books.

Seid, S. (2004) 'Tunneling Through the Wasteland: Ant Farm Video', in C. M. Lewallen and S. Seid (eds), *Ant Farm 1968–1978*, Berkeley, CA: University of California Press / Berkeley Art Museum and Pacific Film Archive, pp. 22–37.

Shank, G. (1993) 'Abductive Multiloguing: The Semiotic Dynamics of Navigating the Net', *The Arachnet Electronic Journal on Virtual Culture*, 1(1), 22 March, <http://www.ibiblio.org/pub/academic/communications/papers/ejvc/SHANK. V1N1> (accessed 27 November 2010).

Sheridan, T. (1992) 'Musings on Telepresence and Virtual Presence', *Presence: Teleoperators and Virtual Environments*, 1(1): 120–125.

Shubert, H. (2004) 'Cedric Price: Fun Palace', *Domus*, 866, January: 17.

Sibley, D. (1995) *Geographies of Exclusion: Society and Difference in the West*, London: Routledge.

Silver, A. (1990) 'The Curious Importance of Small Groups in American Sociology', in H. J. Gans (ed.), *Sociology in America*, Newbury Park, CA: Sage, pp. 61–72.

Silverstone, R., and Haddon, L. (1996) 'Design and the Domestication of Information and Communication Technologies: Technical Changes and Everyday Life', in R. Mansell and R. Silverstone (eds), *Communication by Design: The Politics of Information and Communication Technologies*, Oxford: Oxford University Press, pp. 44–74.

Silverstone, R., Hirsch, E., and Morley, D. (1994) 'Information and Communication Technologies and the Moral Economy of the Household', in R. Silverstone and E. Hirsch (eds), *Consuming Technologies: Media and Information in Domestic Spaces*, London: Routledge, pp. 15–31.

Sinclair, J. (2003) '"The Hollywood of Latin America": Miami as Regional Center in Television Trade', *Television and New Media*, 4(3), August: 211–229.

Sinclair, J., and Cunningham, S. (2000) 'Diasporas and the Media', in S. Cunningham and J. Sinclair (eds), *Floating Lives: The Media and Asian Diasporas*, St Lucia, Queensland: University of Queensland Press, pp. 1–34.

Slater, D. (2002) 'Social Relationships and Identity Online and Offline', in L. A. Lievrouw and S. Livingstone (eds), *Handbook of New Media: Social Shaping and Consequences of ICTs*, London: Sage, pp. 533–546.

——. (2003) 'Modernity Under Construction: Building the Internet in Trinidad', in T. J. Misa, P. Brey and A. Feenberg (eds), *Modernity and Technology*, Cambridge, MA: MIT Press, pp. 139–160.

Smith, N. (1993) 'Homeless/Global: Scaling Places', in J. Bird, B. Curtis, T. Putnam, G. Robertson, and L. Tickner (eds), *Mapping the Futures: Local Cultures, Global Change*, London: Routledge, pp. 87–119.

Smith, N., and Katz, C. (1996) 'Grounding Metaphor: Towards a Spatialized Politics', in M. Keith and S. Pile (eds), *Place and the Politics of Identity*, London: Routledge, pp. 67–83.

Sorokin, P. A. (1963) 'Foreword', in F. Tönnies, *Community and Society (Gemeinschaft und Gesellschaft)*, trans. & ed. C. P. Loomis, New York: Harper and Row, pp. vii–viii.

Spigel, L. (1990) 'Television in the Family Circle: The Popular Reception of a New Medium', in P. Mellencamp (ed.), *Logics of Television: Essays in Cultural Criticism*, Bloomington, IN / London: Indiana University Press / BFI, pp. 73–97.

———. (1992) *Make Room for TV: Television and the Family Ideal in Postwar America*, Chicago: The University of Chicago Press.

———. (2001) 'Media Homes: Then and Now', *International Journal of Cultural Studies*, 4(4): 385–411.

———. (2002) 'Installing the Television Set' in B. Highmore (ed.), *The Everyday Life Reader*, London: Routledge, pp. 325–338.

———. (2004) 'Portable TV: Studies in Domestic Space Travels', in M. Sturken, D. Thomas, and S. J. Ball-Rokeach (eds), *Technological Visions: The Hopes and Fears that Shape New Technologies*, Philadelphia, PA: Temple University Press, pp. 110–144.

———. (2005) 'Designing the Smart House: Posthuman Domesticity and Conspicuous Consumption', *European Journal of Cultural Studies*, 8(4): 403–426.

Spiller, N. (1995) 'Hot Desking in Nanotopia', *Architectural Design* (Architectural Design Profile No. 118: 'Architects in Cyberspace'), 65(11/12), November–December: 70–75.

———. (1998a) *Digital Dreams: Architecture and the New Alchemic Technologies*, London: Ellipsis.

———. (1998b) 'Vacillating Objects', *Architectural Design* (Architectural Design Profile No. 136: 'Architects in Cyberspace II'), 68(11/12), November–December pp. 57–59.

———. (2002) 'Cedric Price_Generator Project_1976', in N. Spiller (ed.), *Cyber_ Reader: Critical Writings for the Digital Era*, London: Phaidon, pp. 84–85.

Spivak, G. C. (1997) 'Translator's Preface', in J. Derrida, *Of Grammatology*, trans. G. C. Spivak, Baltimore, MD: The Johns Hopkins University Press, pp. ix–lxxxvii.

Starrs, P. F., and Anderson, J. (1997) 'The Words of Cyberspace', *Geographical Review*, 87(2), April: 146–154.

Steadman, P. (1979), *The Evolution of Designs: Biological Analogy in Architecture and the Applied Arts*, Cambridge, UK: Cambridge University Press.

Steuer, J. (1992) 'Defining Virtual Reality: Dimensions Determining Presence', *Journal of Communication*, 42(4), Autumn: 73–93.

Stone, A. R. (1992) 'Will the Real Body Please Stand Up?: Boundary Stories About Virtual Cultures', in M. Benedikt (ed.), *Cyberspace: First Steps*, Cambridge, MA: MIT Press, pp. 81–118.

———. (1995) *The War of Desire and Technology at the Close of the Mechanical Age*, Cambridge, MA: MIT Press.

Stone, L. (1991) 'The Public and the Private in the Stately Homes of England, 1500–1990', *Social Research*, 58(1), Spring: 227–251.

Storey, J. (2001) *Cultural Theory and Popular Culture: An Introduction*, 3rd edition, Harlow, UK: Prentice Hall.

Strate, L. (1999) 'The Varieties of Cyberspace: Problems in Definition and Delimitation', *Western Journal of Communication*, 63(3), Summer: 382–412.

Sutherland, I. E. (1963) 'SKETCHPAD: A Man-Machine Graphical Communication System', *AFIPS Conference Proceedings*: 329–346.

———. (2003) 'Sketchpad: A Man-Machine Graphical Communication System, Technical Report No. 574', Cambridge, UK: Computer Laboratory, University of Cambridge.

Tabor, P. (1998) 'Striking Home: The Telematic Assault on Identity', in J. Hill (ed.), *Occupying Architecture: Between the Architect and the User*, London: Routledge, pp. 217–228.

Tække, J. (2002) 'Cyberspace as a Space Parallel to Geographical Space', in L. Qvortrup (ed.), *Virtual Space: Spatiality in Virtual Inhabited 3D Worlds*, London: Springer, pp. 25–46.

Telleen, S. (1998) 'What It Means to Have Virtual Communities on an Intranet', *Internet World*, 16 November: 21.

Theall, D. (1995) *Beyond the Word: Reconstructing Sense in the Joyce Era of Technology, Culture, and Communication*, Toronto: University of Toronto Press.

Thien, D. (2005) 'After or Beyond Feeling? A Consideration of Affect and Emotion in Geography', *Area*, 37(4), December: 450–456.

'This Is the Toaster Telling You I'm On Fire' (1995), *Age: Computer Age*, 25 April: 1 & 7.

Thomas, J., and Warren, B. (eds) (2000) *Proceedings of the First Australasian User Interface Conference (AUIC 2000): Stepping out of Windows*, Los Alamitos, CA: IEEE Computer Society.

Thorns, D. C. (1976) *The Quest for Community: Social Aspects of Residential Growth*, London: George Allen & Unwin Ltd.

Thrift, N. (1999) 'The Place of Complexity', *Theory, Culture & Society*, 16(3): 31–69.

———. (2004) 'Intensities of Feeling: Towards a Spatial Politics of Affect', *Geografiska annaler*, 86B(1): 57–78.

Todorov, T. (1996) 'Living Alone Together', *New Literary History*, 27(1), Winter: 1–14.

Tofts, D. (1999) 'Hyperlogic, the Avant-Garde and Other Intransitive Acts', in *Parallax: Essays on Art, Culture and Technology*, North Ryde, Sydney, NSW: Interface, pp. 16–28.

———. (2000a) '*E-Topia: "Urban Life, Jim—But Not as We Know It"*', *Screening the Past*, 9, <http://www.latrobe.edu.au/screeningthepast/shorts/reviews/rev0300/dtbr9a.htm> (accessed 27 November 2010).

———. (2000b) 'Terrible Beauty: The Work of Murray McKeich', *Photofile: Photomedia Journal* ('Tekhne'), 60, August: 12–16.

———. (2004a) 'Medea Theory', *Rhizomes*, 9, Fall, <http://www.rhizomes.net/issue9/tofts.htm> (accessed 27 November 2010).

———. (2004b) 'Unexpected Innovations', *RealTime*: OnScreen / Digital, 63, October–November: 32.

———. (2005) 'f2f 2 url & b ond: Space/Time and the Dissemination of Community', *Transformations*, 12, December, http://transformations.cqu.edu.au/journal/issue_12/article_01_print.shtml (accessed 16 December 2005).

Tofts, D., Jonson, A., and Cavallaro, A. (eds) (2002) *Prefiguring Cyberculture: An Intellectual History*, Sydney / Cambridge, MA: Power Publications / MIT Press.

Tofts, D., and McKeich, M. (1998) *Memory Trade: A Prehistory of Cyberculture*, North Ryde, Sydney, NSW: 21C / Interface.

Tönnies, F. (1963) *Community and Society (Gemeinschaft und Gesellschaft)*, trans. C. P. Loomis, New York: Harper Row.

Tuan, Y.-F. (1977) *Space and Place: The Perspective of Experience*, Minneapolis, MN: University of Minnesota Press.

Turkle, S. (1995) *Life on the Screen: Identity in the Age of the Internet*, New York: Simon and Schuster.

Turner, B. (2005) 'Obituaries and the Legacy of Derrida', *Theory, Culture & Society*, 22(2): 131–136.

Turner, F. (2006) *From Counterculture to Cyberculture: Stewart Brand, the Whole Earth Network, and the Rise of Digital Utopianism*, Chicago, IL: The University of Chicago Press.

Ulmer, G. L. (1988) 'The Puncept in Grammatology', in J. Culler (ed.), *On Puns: The Foundation of Letters*, Oxford: Basil Blackwell, pp. 164–189.

———. (1994) *Heuretics: The Logic of Invention*, Baltimore, MD: The Johns Hopkins University Press.

Uncapher, W. (1999) 'Electronic Homesteading on the Rural Frontier: Big Sky Telegraph and Its Community', in M. A. Smith and P. Kollock (eds), *Communities in Cyberspace*, London: Routledge, pp. 264–289.

'The Un-Private House' (1999), Museum of Modern Art, New York, exhibition, <http://www.moma.org/exhibitions/1999/un-privatehouse/index.html> (accessed 27 November 2010).

Urry, J. (2000a) 'Mobile Sociology', *British Journal of Sociology*, 51(1), January/March: 185–203.

———. (2000b) *Sociology Beyond Societies: Mobilities for the Twenty-First Century*, London: Routledge.

———. (2002) 'Mobility and Proximity', *Sociology*, 36(2): 255–274.

———. (2005) 'The Complexity Turn', *Theory, Culture & Society*, 22(5): 1–14.

Van Den Abbeele, G. (1991) 'Introduction', in Miami Theory Collective (ed.), *Community at Loose Ends*, Minneapolis, MN: University of Minnesota Press, pp. ix–xxvi.

———. (1997) 'Lost Horizons and Uncommon Grounds: For a Poetics of Finitude in the Work of Jean-Luc Nancy', in D. Sheppard, S. Sparks, and C. Thomas (eds), *On Jean-Luc Nancy: The Sense of Philosophy*, London: Routledge, pp. 12–18.

VanEvery, J. (1999) 'From Modern Nuclear Family Households to Postmodern Diversity?: The Sociological Construction of "Families"', in G. Jagger and C. Wright (eds), *Changing Family Values*, London: Routledge, pp. 165–184.

Vasseleu, C. (1997) 'Virtual Bodies/Virtual Worlds', in D. Holmes (ed.), *Virtual Politics: Identity and Community in Cyberspace*, Thousand Oaks: Sage, pp. 46–58.

Vidich, A. J., and Bensman, J. (1968a) 'Small Town in Mass Society', in R. L. Warren (ed.), *Perspectives on the American Community: A Book of Readings*, 4th printing, Chicago: Rand McNally & Company, pp. 201–213.

———. (1968b) *Small Town in Mass Society: Class, Power and Religion in a Rural Community*, revised edition, Princeton, NJ: Princeton University Press.

Vidler, A. (1990) 'The Building in Pain: The Body and Architecture in Post-Modern Culture', *AA Files*, 19, Spring: 3–10.

———. (1992) *The Architectural Uncanny: Essays in the Modern Unhomely*, Cambridge, MA: MIT Press.

Virilio, P. (1998) 'We May Be Entering an Electronic Gothic Era', *Architectural Design* (Architectural Design Profile No. 136: 'Architects in Cyberspace II'), 68(11/12), November–December: 61.

———. (2001) 'Landscape of Events Seen at Speed: Interview with Pierre Sterckx', in J. Armitage (ed.), *Virilio Live: Selected Interviews*, London: Sage, pp. 144–153.

Voller, J. G. (1993) 'Neuromanticism: Cyberspace and the Sublime', *Extrapolation*, 34(1), Spring: 18–29.

Von Bertalanffy, L. (1950) 'An Outline of General Systems Theory', *The British Journal for the Philosophy of Science*, 1(1), May: 134–165.

———. (1951) 'General System Theory: A New Approach to Unity of Science', *Human Biology: A Record of Research*, 23(4), December: 346–361.

———. (1960) *Problems of Life: An Evaluation of Modern Biological and Scientific Thought*, New York: Harper Torchbooks.

———. (1967) *Robots, Men and Minds: Psychology in the Modern World*, New York: George Braziller.

———. (1968) *General System Theory: Foundations, Development, Applications*, New York: George Braziller.

Von Schelling, F. W. J. (1989) *The Philosophy of Art*, trans. D. W. Stott, Minneapolis, MN: University of Minnesota Press.

Ward, K. (1998) 'The Emergence of the Hybrid Community: Re-Thinking the Virtual/Physical Dichotomy', *Space and Culture*, 1(4–5): 71–86.

Warf, B., and Grimes, J. (1997) 'Counterhegemonic Discourses and the Internet', *Geographical Review*, 87(2), April: 259–274.

Wark, M. (1990) *Virtual Geography: Living with Global Media Events*, Bloomington, IN: Indiana University Press.

———. (1993) 'Suck on This, Planet of Noise! (Version 1.2)', in D. Bennett (ed.), *Cultural Studies: Pluralism and Theory*, Melbourne: Melbourne University Press, pp. 156–170.

———. (1994) 'Third Nature', *Cultural Studies*, 8(1), January: 115–132.

———. (1997) 'Cyberhype: The Packaging of Cyberpunk', in A. Crawford and R. Edgar (eds), *Transit Lounge*, North Ryde, Sydney, NSW: 21C / Interface, pp. 154–157.

Warren, R. L. (1968) 'Section Three: The Community's Vertical and Horizontal Patterns—Introduction', in R. L. Warren (ed.), *Perspectives on the American Community: A Book of Readings*, 4th printing, Chicago, IL: Rand McNally & Company, pp. 193–200.

Waters, M. (1995) *Globalization*, London: Routledge.

Watson, N. (1997) 'Why We Argue About Virtual Community: A Case Study of the Phish.Net Fan Community', in S. G. Jones (ed.), *Virtual Culture: Identity and Communication in Cybersociety*, London: Sage, pp. 102–132.

Wei, R., and Lo, V.-H. (2006) 'Staying Connected While on the Move: Cell Phone Use and Social Connectedness', *New Media and Society*, 8(1): 53–72.

Weimann, G. (2000) *Communicating Unreality: Modern Media and the Reconstruction of Reality*, Thousand Oaks: Sage.

Weiss, P. A. (1995) 'Feminist Reflections on Community', in P. A. Weiss and M. Friedman (eds), *Feminism and Community*, Philadelphia, PA: Temple University Press, pp. 3–18.

Weiss, P. A., and Friedman, M. (eds) (1995) *Feminism and Community*, Philadelphia, PA: Temple University Press.

Wellesley-Miller, S. (1975) 'Intelligent Environments', in N. Negroponte, *Soft Architecture Machines*, Cambridge, MA: MIT Press, pp. 125–129.

Wellman, B. (2001) 'Physical Place and Cyberplace: The Rise of Networked Individualism', in L. Keeble and B. Loader (eds), *Community Informatics: Shaping Computer-Mediated Social Relations*, London: Routledge, pp. 17–42.

Wellman, B., and Gulia, M. (1999) 'Virtual Communities as Communities: Net Surfers Don't Ride Alone', in M. A. Smith and P. Kollock (eds), *Communities in Cyberspace*, London: Routledge, pp. 167–194.

Werner, C. M. (1987) 'Home Interiors: A Time and Place for Interpersonal Relationships', *Environment and Behavior*, 19(2), March: 169–179.

Werner, F. (2001) *Covering + Exposing: The Architecture of Coop Himmelb(l)au*, trans. M. Robinson, Basel, Switzerland: Birkhäuser Verlag AG.

Wertheim, M. (1999) *The Pearly Gates of Cyberspace: A History of Space from Dante to the Internet*, Sydney: Doubleday.

Wheelock, J. (1994) 'Personal Computers, Gender and an Institutional Model of the Household', in R. Silverstone and E. Hirsch (eds), *Consuming Technologies: Media and Information in Domestic Spaces*, London: Routledge, pp. 97–112.

White, M. (2004) 'Re: First Post (an Internet without Space)', online posting to air-l@aoir.org.list, 3 February (accessed 6 February 2004).

White, N. (2005) 'How Some Folks Have Tried to Describe Community', April, <http://www.fullcirc.com/community/definingcommunity.htm> (accessed 27 November 2010).

Whittle, D. B. (1997) *Cyberspace: The Human Dimension*, New York: W. H. Freeman.

Wigley, M. (1998) *Constant's New Babylon: The Hyper-Architecture of Desire*, Rotterdam: Witte de With, Center for Contemporary Art / 010 Publishers.

———. (2002) *The Architecture of Deconstruction: Derrida's Haunt*, 6th printing. Cambridge, MA: MIT Press.

———. (2004) 'Anti-Buildings and Anti-Architects', *Domus*, 866, January: 15–16, 22.

———. (2006) 'Network Fever', in W. H. K. Chun and T. Keenan (eds), *New Media, Old Media: A History and Theory Reader*, London: Routledge, pp. 375–397.

Wilbur, S. (2000) 'An Archaeology of Cyberspaces: Virtuality, Community, Identity', in D. Bell and B. Kennedy (eds), *Cybercultures Reader*, London: Routledge, 45–55.

Wilken, R. (2008) 'Mobilizing Place: Mobile Media, Peripatetics, and the Renegotiation of Urban Places', *Journal of Urban Technology*, 15(3): 39–55.

———. (2010) 'A Community of Strangers? Mobile Media, Art, Tactility and Urban Encounters with the Other', *Mobilities*, 5(4), November: 449–468.

Wilken, R., and Sinclair, J. (2011) 'Global Marketing Communications and Strategic Regionalism', *Globalizations*, 8(1) February: 1–15.

Williams, J. (2003) *Gilles Deleuze's* Difference and Repetition: *A Critical Introduction and Guide*, Edinburgh: Edinburgh University Press.

Williams, R. (1976) *Keywords: A Vocabulary of Culture and Society*, London: Fontana / Croom Helm.

———. (1989) *Raymond Williams on Television: Selected Writings*, A. O'Connor (ed.), London: Routledge.

———. (1992) *Television: Technology and Cultural Form*, Hanover: Wesleyan University Press / University Press of New England.

Williams, R. L., and Cothrel, J. (2000) 'Four Smart Ways To Run Online Communities', *Sloan Management Review*, Summer: 81–91.

Willson, M. (1997) 'Community in the Abstract: A Political and Ethical Dilemma?', in D. Holmes (ed.), *Virtual Politics: Identity and Community in Cyberspace*, Thousand Oaks: Sage, pp. 145–162.

———. (2006) *Technically Together: Rethinking Community within Techno-Society*, New York: Peter Lang.

Wolmark, J. (ed.) (1999) *Cybersexualities: A Reader on Feminist Theory, Cyborgs and Cyberspace*, Edinburgh: Edinburgh University Press.

Wood, J. (ed.) (1998) *The Virtual Embodied: Presence / Practice / Technology*, London: Routledge.

Woods, L. (1996) 'The Question of Space', in S. Aronowitz, B. Martinsons, M. Menser, and J. Rich (eds), *Technoscience and Cyberculture*, London: Routledge, pp. 279–292.

Woodward, I. (2001) 'Domestic Objects and the Taste Epiphany: A Resource for Consumption Methodology', *Journal of Material Culture*, 6(2): 115–136.

Wrede, S. (ed.) (1990), *Architectural Drawings of the Russian Avant-Garde*, New York: Museum of Modern Art.

Wright, F. L. (1970) *An Organic Architecture: The Architecture of Democracy*, Bradford: Lund Humphries.

Wright, G. (1991) 'Prescribing the Model Home', *Social Research*, 58(1), Spring: 213–225.

Yates, F. (1996) *The Art of Memory*, London: Pimlico.
Yoon, K. (2003) 'Retraditionalizing the Mobile: Young People's Sociality and Mobile Phone Use in Seoul, South Korea', *European Journal of Cultural Studies*, 6(3): 327–343.
Young, I. M. (1995) 'The Ideal of Community and the Politics of Difference', in P. A. Weiss and M. Friedman (eds), *Feminism and Community*, Philadelphia, PA: Temple University Press, pp. 232–257.
Zamyatin, Y. (1972) *We*, trans. B. G. Guerney, Harmondsworth, Middlesex: Penguin.
Zuk, W., and Clark, R. H. (1970) *Kinetic Architecture*, New York: Van Nostrand Reinhold Company.

Index

For Product Safety Concerns and Information please contact our EU
representative GPSR@taylorandfrancis.com
Taylor & Francis Verlag GmbH, Kaufingerstraße 24, 80331 München, Germany